Hobby Farmer's Handbook

TO RAISING GOATS

I0458788

KAREN HARRIS

Publication Data

Karen Harris
Hobby Farmer's Handbook to Raising Goats – First edition.
Summary: "Your complete guide to starting and managing a thriving goat herd"
Provided by publisher.
ISBN: 978-1-961846-10-4
[1. Hobby Farmer's Handbook to Raising Goats – Non-Fiction] I. Title.

This book has been written with the published intent to provide accurate and authoritative information in regard to the subject matter included. While every reasonable precaution has been taken in preparation of this book the author and publisher expressly disclaim responsibility for any errors, omissions, or adverse effects arising from the use or application of the information contained inside. The techniques and suggestions are to be used at the reader's discretion and are not to be considered a substitute for professional veterinary care. If you suspect a medical problem with your goat, consult your veterinarian.

Design by Sorin Rădulescu
First paperback edition, 2024

TABLE OF CONTENTS

Chapter 7

Goat Nutrition and Feeding

Chapter 8

Troubleshooting and Problem-solving

Chapter 9

Chapter 10

Chapter 11

Chapter 12

CHAPTER 1

Introduction to Goat Farming

Almost as soon as we signed the papers to purchase our five-acre hobby farm, my husband began making plans to get goats. "What good is a goat?" I remember asking him. The answer to my question, it turns out, was not as short and simple as I thought. There are numerous reasons why hobby farmers—small-scale or backyard farmers—should consider adding a few goats to their agricultural activities. I may have entered the world of small-herd goat farming as a skeptic, but the cuteness of goats and their hilarious personalities quickly won me over.

Several years ago, when the kids were younger, they showed goats at our county fair as part of 4-H, so that was our excuse for raising goats on our hobby farm. Yet, as the kids phased out of 4-H, we continued to keep goats. We found them to be enjoyable and fairly easy to care for. The goats have morphed into companion animals, like pets, but ones that don't sleep at the end of my bed!

Because we have just about five acres of land, our hobby farming activities are done on a smaller scale as compared to large-scale commercial farms. We have never had more than

HISTORICAL FACT
Ancient Origins

Goats were among the first domesticated animals, with over 10,000 years of history. Early domestic goats played a critical role for Neolithic farmers, providing vital nutrition in the form of both meat and dairy. In addition to food products, these farmers likely used goats' skin and hair for clothing, their bones for tools, and possibly their dung for fuel. The domestication of goats originated in several distinct communities in the Middle East and Western Asia.

Photo Courtesy of April Wilson

five goats at a time, and we have not derived an income from our animals. For us, goats enhance the rural lifestyle we have tried to achieve. Yet that doesn't mean that small-scale goat farmers can't profit from their animals. As we will see throughout this book, it is totally possible. I know plenty of folks who do.

In the coming chapters, we will delve into the care and keeping of goats and offer a guide to managing a healthy and thriving small-scale goat herd. You will find information on the various breeds of goats and tips on selecting the breed that best suits your goals and situations. You'll learn about the housing, pasture, and feed requirements of goats, as well as preventative measures to keep them healthy. We will offer a basic guide to some of the most common goat diseases, along with symptoms and treatments for each of them. We have included an overview of the goat breeding and kidding process, as well as a chapter on milking your goats, safe milk handling, and uses for goat milk. You will also find a chapter on raising meat goats and meat processing. Throughout the *Hobby Farmer's Handbook to Raising Goats*, we will address common challenges to small-scale goat farming and discuss solutions. I will even regale you with a few tales from my goat-raising adventures (or misadventures!) and introduce you to a few of the goats that have captured my heart, like Juju, Brown Sugar (I call her Suge), and Dixie.

Before we launch into our guide to small-scale, hobby-farm goat raising, let me return to the question I originally asked my husband when he announced that he wanted to get goats so that I can properly answer it now, years later and with the benefit of experience.

What Good Is a Goat?

Goats, I have learned, are incredibly versatile animals. They are more than just livestock with personalities or 4-H projects for the kids.

Goats and humans, in fact, have a long history of symbiotic relations stretching back more than 10,000 years. Humans first domesticated goats in the region known as the Fertile Crescent, which encompasses parts of present-day Iran, Iraq, and Syria, a place that anthropologists often refer to as the cradle of civilization. It was here, after all, that our ancient ancestors first dabbled in farming practices and the domestication of animals for livestock. It is believed, in fact, that the early domestication of goats played an important role in the development of human societies, as the animals offered a reliable source of food.

Raising Goats for Milk

One of the primary reasons why people raised goats in antiquity, as well as today, is for their milk. In the United States, most people drink cow's milk; however, goat's milk is more popular worldwide. More people across the globe consume goat's milk—and products made from goat's milk—than cow's milk. For many people, goat milk is easier to digest. It

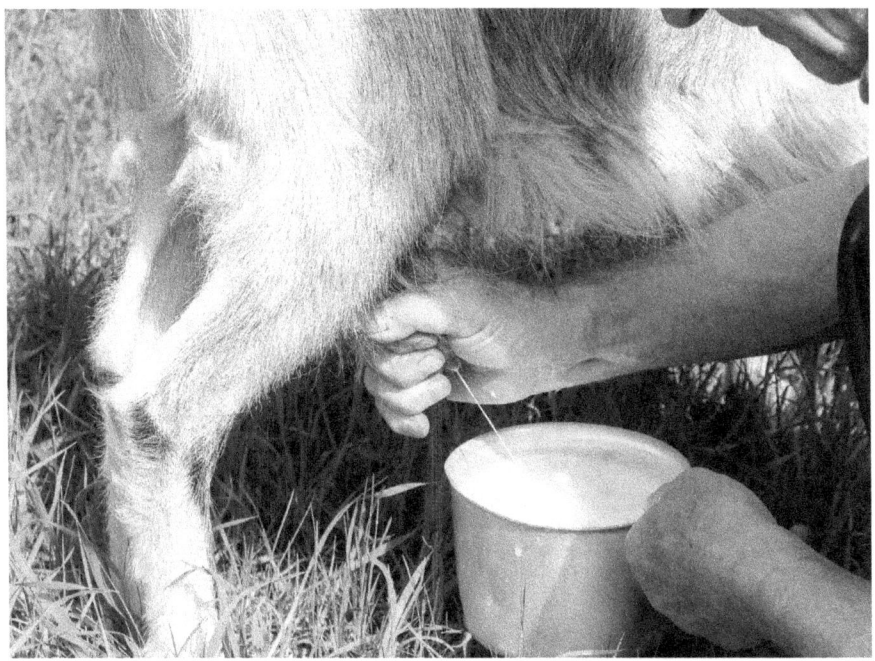

has a lower percentage of lactose and is naturally homogenized. We will delve into this more in Chapter 9.

Goat's milk is also popular for cheese and soapmaking, two activities that dovetail nicely into the hobby-farm lifestyle. Some breeds of goats, particularly the angora goat, produce wool that is commonly used in specialty textiles.

Raising Goats for Meat

Goats can also be raised for meat. Although goat meat is not as popular in the United States, Canada, and throughout Europe, it is the go-to red meat in many parts of the world. Goat consumption in the U.S. is on the rise thanks to an increased interest in global cuisine. As you will learn in Chapter 10, goat meat has a reputation for being the healthiest of all red meats. It is lower in fat and calories, too. Goat meat is even healthier than chicken.

Raising Goats for 4-H

If, like us, you want your children to be involved in 4-H—and I would recommend it. It is, after all, the country's largest youth development organization—goats make a great project, especially for younger kids and families just starting out in 4-H. They are smaller, cheaper, and easier to handle than other livestock animals, like dairy cows and pigs. Participating in goat shows at county fairs is a fun activity that involves the whole family. Although my kids were never grand champions at our fair's goat shows, they did learn about responsibility, and they met some new friends.

Raising Goats for Land Management

On a recent trip to Biltmore Mansion, America's largest home, located in Asheville, North Carolina, I was delighted to see a pen of goats on the grounds of the sprawling estate. A nearby sign informed visitors that it was a "working herd." The goats were periodically called on for land management duties. You might not be a Vanderbilt, but you, too, can consider getting goats for land management purposes. As natural-born browsers, goats will clear out overgrown areas and control invasive plant species. It is true—goats are like lawnmowers and will keep your pasture neatly trimmed. The virtues of brush-control goats are becoming more well-known. You have probably seen reports on the news from time to time about how a herd of goats was "hired" to clear out the underbrush at a state or national park as a sustainable land management technique.

Raising Goats for Fiber

Some varieties of goats produce fibers that are utilized in the textile industry. There is a growing interest in homespun cloth, and it is not uncommon to see textile-related cottage industries as part of a hobby farm that raises goats for cashmere and mohair. In fact, these fibers are a secondary product, as goats are primarily used for either milk or meat. Raising goats for their fibers just adds another layer of value to the animals and another potential revenue stream for a small-scale hobby farm.

Raising Goats as Pack Animals

Goats are also useful as pack animals. Although they cannot carry loads as heavy as mules, donkeys, or horses, the surefootedness of goats makes them well-suited for carrying lighter loads over rugged, mountainous terrains. Goats can be trained to pull carts or small wagons, but again, they are not as strong as other pack animals. If your hobby farm is hilly, rocky, and rugged, goats can be used to transport items around your farm, even in places that are inaccessible by wheel.

Raising Goats as Companion Pets

Lastly, goats are good companions. They are quite trainable and extremely intelligent. Each one has its own unique personality and funny quirks. Many hobby farmers find great satisfaction in building relationships with their goats and take delight in spending time with them.

Are Goats Right For You?

Before you get started on your adventures in small-scale goat farming, you need to ask yourself a few more questions, beyond my "What good is a goat?" Consider the following questions to help you determine if goats are a good choice for your hobby farm and to narrow down what you hope to get out of your goat-raising endeavors.

Can You Legally Have a Goat Herd?

You might think, "It is my land. I can do what I want on it." Unfortunately, that is not always the case. Even though you have purchased your land, you are still required to abide by local zoning laws, regulations, and ordinances. These may be the deciding factors in determining whether you are able to raise goats or not. The first thing you need to do is contact your local zoning authority to see if your home is in a district zoned for small-scale agricultural operations.

Be clear when you ask your questions and inform the zoning official that your plan is to raise a small herd of goats. If you ask if your property is zoned for agricultural use, the official may assume you mean larger livestock animals, like cattle or hogs. Small-scale hobby farming is such a growing trend in the U.S. that, in many communities, small herds are permitted in residential areas if the property is over a specified size, like one acre.

Ask if you are able to earn an income from your goats. In some places, the local laws and ordinances prohibit engaging in agricultural activities for profit, meaning you cannot sell goat cheese or mohair sweaters. The laws might be worded to say that only livestock animals for personal use are allowed.

Your town or community might also have ordinances prohibiting "nuisance" animals—animals that are likely to annoy the neighbors. Roosters and peacocks fall under this label, but occasionally, so do goats.

Your county or town may give goats the green light, but if you live in a neighborhood governed by a homeowner's association, you will have to check their rules as well. Their laws supersede local ordinances.

In many places, hobby farms are legally permitted to raise goats if they abide by a set of rules. These may include the number of goats (some say no less and no more than two goats) and the size of the goats (for example, only pygmy or dwarf goats are allowed). There might be specifications that all goats be dehorned and that all male goats must be castrated. Expect to also find requirements regarding the housing and outdoor enclosure. The size, placement, fence height, and building materials may be spelled out for you, as well as how you are to handle the removal of manure.

Contact your county extension office to find out if you need a permit to raise goats in your county or state. The extension officer can also tell you if you need to register your animals, if they need to be tested for communicable diseases, and if there are additional requirements. In my state, for example, goats have a scrapie tag or tattoo issued by the U.S. Department of Agriculture. If I wanted to take my goats to another state, to enter them in a livestock show, for example, or to sell them to an out-of-state buyer, each goat would need to have an official interstate certificate of veterinary inspection, also called a CVI, given to me by a USDA-accredited veterinarian.

Are Goats Covered under Your Homeowner's Insurance?

While we are on the subject of the legalities of small-scale goat farming, let's take a moment to talk about your insurance coverage. Goats, as we will learn throughout this book, are highly skilled escape artists. What would happen if, for example, one of your goats got loose and ate your neighbor's flower gardens? Would your homeowner's insurance cover the cost of replacing the damaged landscaping plants or any other damage that your goats might cause? That all depends on your policy.

Protect yourself from potential risks by calling your insurance agent to find out what is and is not covered under your current policy. If goat damage falls under the "not covered" category, ask if there is an additional insurance rider you can purchase that will protect you. Most insurance companies offer riders for hobby farm or agricultural claims.

When you have your insurance agent on the phone, also ask about coverage on your goat barn and pasture. I speak from experience here. The morning after our first goat barn was flattened by a downed tree, I called to file a claim on it. It was only then that I found out that when our homeowner's policy stated that "outbuildings" were included, that didn't mean buildings used for housing agricultural animals. Outbuildings, to the insurance company, meant detached garages, storage sheds, and the like. Since we didn't have a rider to cover agricultural buildings, we had to pay to replace the goat barn ourselves. Live and learn. We now have the correct rider so that our goat barn and chicken coop are insured. I would strongly suggest you do this as well.

What Do You Want from Your Goats?

Think about what you want to get out of your goat herd. This will help you determine the type of goats you get—dairy goats, meat goats, or perhaps dual-purpose goats. We will look at the most common goat breeds for hobby farmers in Chapter 2 and discuss the qualities of each one.

Do You Have Room for Goats?

You also need to determine if your hobby farm is conducive to goat farming. Do you have enough space? Will you need to build a goat barn? How much pasture space do you need? In Chapter 3, we will answer these questions and give you the information you need to set up your goat area.

Goats need companionship. As pack animals, they are hardwired to live in groups. Raising just one goat is not advisable. The animal will be so lonely and depressed that its health will suffer. Later, in Chapter 3, when we discuss the space requirements for goats, keep in mind that you will need to double that to accommodate a minimum of two goats.

What Will the Neighbors Think?

You also need to consider your neighbors and your community. How will your small-scale goat farm impact others living near you? It is the neighborly thing to do to discuss your intentions with the people next door and listen to their concerns. If you have done your homework, you should be able to put them at ease and address their worries. Explain to

them how you plan to prevent your goats from escaping, how you will employ good hygiene practices to keep the smell down, and how you will handle any other problems that may arise. Ask your neighbors to kindly reach out to you if they are inconvenienced by your goats so that you can work with them to solve the issue. Express how you wish to keep the lines of communication open so there will be no hard feelings or resentment.

Goat Jargon

There is one more topic we need to cover before you are ready to become a small-scale, hobby goat farmer—vocabulary words. If you want to be a goat farmer, you need to sound like a goat farmer. Using the correct terminology will help you communicate with veteran goat enthusiasts and your veterinarian and help you understand some of the information that will come later in this book. Here are a few words you should know:

Bloat – A condition in which the goat's stomach is so full of gas that the abdomen appears dissented.

Browsing – Goats are not true grazers but browsers. They will search out and eat the leaves of trees, woody shrubs, and branches and will only eat grass if there's no bushy vegetation around.

Buck/Billy – A mature, intact male goat used for breeding.

Buckling – A sexually immature, young male goat.

Butting – Goats fight by ramming horns or heads with their opponents.

Caprine – The scientific term for a goat.

Chevron – Either a French term for a goat slaughtered just after weaning or a general term for goat meat.

Doe – A mature female goat.

Doeling – A sexually immature, young female goat.

Freshen – When a doe begins to produce milk because she has given birth.

Kid – Either a young goat that is less than one year old or the act of a doe giving birth.

Polled – A naturally hornless goat.

Rudiments – A group of animals that have four-sectioned stomachs, including goats, cattle, deer, sheep, bison, and elk.

Scurs – Small, spiky protrusions of horn material that grows either on a polled or dehorned goat.

Udder – The goat's mammary gland that produces milk.

Wattle – A fleshy mass hanging from the throat area of a goat.

Wether – A castrated male goat.

Yearling – A goat, either male or female, between one and two years old.

Summary

As we continue through this guidebook, you will discover the advantages of raising goats on a small scale. In fact, goats seem to be ideally suited for hobby farms. As we discovered, a few goats fit quite nicely on an acre of land.

Raising goats is a hobby for the whole family. I was one proud parent when one of my daughters woke up at 5:00 a.m. every day for six weeks so she could mix formula and bottle feed two baby goats before she got ready for school and then rushed home after school to bottle feed them again. For a 14-year-old girl who likes her sleep, this was a huge achievement. She learned some valuable lessons about responsibility, commitment, and sacrifice that she wouldn't have otherwise learned.

Goats are a good introduction to hobby farming because they are less expensive to raise than horses or cows, don't need as much space, and are low-maintenance animals. Goats enjoy the company of people. They will be thrilled to see you when you visit the goat barn and follow you around when you work in the pasture. Comical, amusing, and silly, goats have big personalities and aren't shy about showing their quirky sides.

As a hobby farmer, you are looking for ways to engage in agricultural activities for personal enjoyment and to develop a sustainable lifestyle. You may even be considering goats as a source of supplemental income. Admittedly, I have never profited from my goats, but that doesn't mean it can't be done. My primary motivation for hobby farming is to enjoy the simple pleasures of the past and to reconnect with the type of lifestyle my grandparents had.

For me and my family, raising goats on a small scale has been a rewarding experience. There have been a few misadventures along the way—ask me what happened to the pear tree saplings I lovingly planted in hopes of starting a little orchard—but these make good stories. You'll have your own entertaining stories to tell after you welcome goats to your hobby farm.

The goal of this book is to help guide you through your goat adventures, from selecting the right breed to setting up your goat barn and pasture to ensuring that your small-scale goat herd gets the proper nutrition and health care to keep your hobby farm goat herd healthy and thriving.

CHAPTER 2

Choosing the Right Goat Breeds

Once you have made the decision to add a small goat herd to your hobby farm, the next step is to determine the right goat breed for you and your situation. Choosing the right breed depends on several factors, including your goals for your hobby farm, your level of experience with goats, the resources you have available, and, of course, your personal preferences.

In this chapter, we will look at the different types of goats and highlight the most common goat breeds for small-scale hobby farmers. We will break down the purposes and traits of each breed. Armed with this information, you will be better able to choose the breed of goat that best fits your hobby farm.

Your Goat Goals

To expand on some information from Chapter 1, small-scale hobby farmers should have a plan or goal for their goats. With that goal in mind, you will be better

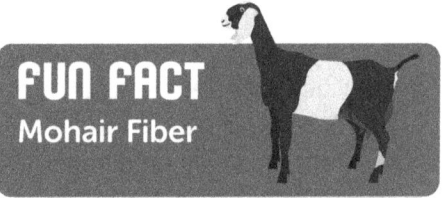

FUN FACT
Mohair Fiber

Mohair is a material produced from the hair of an Angora goat. This luxury fiber is easily dyed and adds luster and texture to knitted garments, making it a popular element of garment construction. Other benefits of this fiber include its water-repellent nature and ability to help regulate body temperature. Most of the world's mohair comes from South Africa, though Angora goats are originally from Turkey. To protect their valuable wool during the winter, some farmers also equip their Angora goats with jackets. An angora goat can produce around 8 to 16 pounds of wool when sheared each year.

equipped to narrow down your goat breed selection. Are you planning to raise goats for milk, meat, fiber, or something else? There are some dual-purpose goat breeds; however, the majority of goat breeds have been developed to accomplish specific purposes and are, therefore, better suited for one goal over another.

Your Previous Goat Experience

Are you new to goat farming, or do you have prior experience handling goats? Some goat breeds can be stubborn, destructive, and ornery, while others are a sheer delight. As a new, inexperienced goat farmer, you could save yourself some grief and stress by choosing a breed of goat with a more docile, tame personality. If you are an experienced goat farmer, however, you probably have the ability to handle a high-strung goat.

Your Resources

The different goat breeds each have different requirements, so you should assess your property and resources to make sure that your farm is equipped to accommodate the goat breed of your choice. Some goat breeds, for example, require more pasture space and taller fencing than others. If you welcome goats to your farm and you lack the resources to appropriately accommodate them, you will have to scramble to make improvements or adjustments to meet the needs of your small goat herd.

Your Personal Preferences

Perhaps there is a certain breed of goat that has caught your eye—one that you think looks adorable or has the qualities you want. By all means, get that goat! Your personal preferences should be considered when selecting a goat breed.

Dairy Goat Breeds

The number one reason why hobby farmers want to raise goats is for their milk. Dairy goats provide a sustainable source of fresh milk, either for drinking or for making dairy products, such as yogurt, cheese, and soap. In general, dairy goats are friendly and sociable. They love people and will bond with their humans.

There are several breeds of dairy goats, but for small-scale hobby farmers, these are the most common ones:

Nubian

Nubians have been our go-to goat breed on our hobby farm, although we have had several different breeds. One of my favorite goats, Brown Sugar, is a Nubian. Like most Nubians, she is gentle and easygoing, with

Nubian

a cheerful personality. This is a talkative goat breed. Nubians have a distinctive vocalization that is often called a "Nubian cry." When you speak to Nubian goats, they will speak back—and they will do so in such an expressive way that you will swear they are trying to communicate with you.

Known for their high milk production, Nubians are an excellent choice for hobby farmers interested in milking goats. This breed produces milk with a high butterfat content, which makes it well-suited for making cheese and butter.

As you can guess from her name, Brown Sugar is a brown Nubian goat with a solid coat. This breed of goat, however, comes in a variety of colors and patterns. We have had several white and brown Nubians, as well as a few black, black and white, and tan and white ones. If you are not sure of the breed of a goat, look at its ears. The ears of Nubian goats are their most recognizable feature. They are long, pendulous, and floppy-looking and resemble large leaves in that they are narrow at the head, widen out in the center, and narrow to a rounded tip. Nubians look like they are wearing their hair in pigtails on either side of their faces.

Nubians have what is called a "Roman profile" or an aquiline nose. Their noses are not sloped or upturned but rather curve outward from a prominent bridge.

Brown Sugar is a big girl. As is typical of Nubian goats, she stands about 32 inches tall at the shoulder and weighs more than 220 pounds. She is rather leggy, though, and appears more delicate than you would think a 200-plus-pound goat would look. When we got our first Nubian, Callie, I was worried that our daughters, who were quite young at the time, wouldn't be able to handle such a tall goat, but I have been proven wrong again and again. Nubians are laid-back and amiable. Brown Sugar is just as sweet as her name.

Alpine

Alpine goats, a breed that hails from the French Alps, are another common dairy goat for hobby farmers because of their many desirable traits. They are excellent milk producers, which is why they are the number one breed for commercial dairy production, as well as popular

Alpine

homestead or hobby-farm goats. The milk of Alpine goats is high in butterfat and can be used to make cheese, ice cream, butter, and soap.

Overall, Alpine goats have a reputation for their calm temperament, making them easier to handle and work with, especially for novice hobby farmers or families with young children. They are gentle and willing to cooperate.

Since this breed originated in the Alps, Alpines are particularly hardy goats that adapt well to various climates. They are robust enough to tolerate extreme cold temperatures but can also handle hot weather. For this reason, Alpines appeal to hobby farmers living in areas where they experience all four seasons.

Alpines, like Nubians, come in a variety of colors and markings, such as solid white, solid black, various shades of brown, and combinations of all of these. It is easy to tell an Alpine goat from a Nubian one. Alpines have a straight profile and erect ears that are generally smaller than those of a Nubian. Considered a medium-sized goat breed, male Alpines are typically about 32 inches tall at the shoulder and weigh roughly 170 pounds. Females are a few inches shorter and weigh around 135 pounds.

Saanen

Saanen

A Swiss dairy goat, the Saanen used to be a breed that was only found in Switzerland, but in the last century, these goats have been exported to other countries. Today, Saanen goats are found around the world. There is one main reason why this goat has gone global—it is Switzerland's most productive breed of dairy goats. For the Saanen, it is not just about quantity. It is also about quality. Saanen goat milk has a comparatively low butterfat content, making this milk more appealing for drinking.

Saanen goats are large. In fact, they are the largest of all the Swiss goat breeds. They are sturdy and well-built with a straight profile. Male Saanen goats average about 185 pounds and stand nearly 38 inches tall at the shoulder. The females top out at approximately 130 pounds and are about 31 inches tall. The goats' erect ears point up and forward. They have shaggier fur than Nubians and are considered to have medium-length hair.

All Saanen goats are white or cream in color. Not only is their fur white, but their skin is white as well. Saanens can easily sunburn in strong sunlight. They need shady places where they can escape from the sun. Although the sun's rays can be troublesome, the Saanen goats are not bothered by the heat. They can tolerate high temperatures, as well as frigid ones.

If you decide to raise Saanen goats on your hobby farm, you will need a spacious pasture that has large shade trees. Don't try to plant small trees in hopes that they will grow large enough to provide shade—the goats will just eat them. The trees need to be mature, well-established, and large enough that the goats can't reach the lowest branches. Alternatively, you could build open pavilions in your pasture to give your goats some shade. Just be sure the pavilions are strong enough to withstand the activities of curious goats and that they are made using materials that are safe for goats in case they nibble on the pavilion.

Toggenburg

Another Swiss dairy goat, the Toggenburg, is not as prolific a milk producer as the Saanen and is not as popular worldwide; however, small-scale hobby farmers find their distinctive appearance charming. Their milk is high in butterfat and ideal for cheesemaking, as well as for ice cream and yogurt.

Domesticated in the eastern part of Switzerland, Toggenburg goats were one of the first dairy goat breeds to be imported to the United States from Switzerland. Although each individual goat has its own personality, overall, Toggenburgs are inquisitive, comical, and gregarious. Hobby farmers

Toggenburg

that raise Toggenburgs often consider them to be pets rather than livestock animals.

Toggenburgs have fur that is soft gray, but it is the white facial markings that are unusual. The animals have a U-shaped line of white fur running down one side of their faces, curving at the end of their noses to run back up the other side. Most Toggenburgs have sizable beards as well.

Mid-sized goats, Toggenburgs range from 25 to 29 inches tall at the shoulder and weigh between 115 and 150 pounds, depending on age and gender.

LaMancha

LaMancha

Many hobby farmers give LaManchas a double take. That's because this breed of goat has unique ears. Some would argue that they don't have ears at all, but that's not really true. LaManchas have tiny ears, sometimes called gopher ears. We had a few LaManchas in the past. One of them was named Dixie. Visitors to our hobby farm inevitably exclaimed, "What happened to her ears?" I even had one person ask if Dixie's ears had to be amputated because of frostbite. I had to explain that Dixie's ears were quite normal for her breed.

LaManchas are the only dairy goat breed developed in the United States. A breeder in California, Eula Fay Frey, created the breed, which was officially recognized in 1927. By the 1950s, LaManchas were gaining popularity. Part of the reason for this was their high milk production rate and the high butterfat in their milk.

Most goat experts claim that LaManchas are among the sweetest, most docile, and most pleasant of all dairy goat breeds. Dixie, however, was none of these things. We could not keep her confined to the pasture. She constantly escaped and ate my landscaping flowers, stood on top of my SUV, destroyed my pear trees, and visited the neighbors' yards. One time, my husband spent several hundred dollars and an entire weekend "Dixie-proofing" the pasture fencing. He extended all the fence posts, ran an additional row of fencing, and topped it all with barbed wire. When he was done, he came inside and said, "There! No way Dixie will be able to escape now." I gave him a frown and pointed out the window where Dixie was, once again, lounging on top of my car. I can't quote what my husband said in response.

Admittedly, my experience with LaManchas is quite limited. I don't want to judge the entire breed based on my experience with just one goat. Other hobby farmers I know rave about their LaManchas, so I can only assume Dixie was one of a kind.

Nigerian Dwarf

Nigerian Dwarf goats, as you may have guessed from the name, are a small-sized goat breed. Don't let their diminutive size fool you. Nigerian Dwarf goats are great milk goats. Their small size, between 22 and 24 inches tall at the shoulder, makes them a top choice for hobby farmers on smaller farms. The goats are compact and hardy, so they are able to handle different climates and environments.

Nigerian Dwarf goats are prolific milk producers. On average, you can expect to get between one and two quarts of milk per day—some

Nigerian Dwarf

produce even more. The high butterfat content in the milk makes it ideal for making cheese, yogurt, butter, and other dairy products.

Inside their miniature exteriors are oversized personalities. They are entertaining, lively, and quirky. They generally have an easygoing temperament and enjoy human interaction. They get along with children as well, so many hobby-farm families find that Nigerian Dwarf goats are the best fit for them.

Nigerian Dwarf goats have a slightly concave profile and upright ears. You can find them in a variety of colors—brown, black, tan, and gold. They can also be multicolored; many have white markings on their faces.

As smaller goats, Nigerian Dwarfs require less pasture space than larger goats and lower-height fences. Likewise, their housing can be smaller. When compared to other dairy goats, Nigerian Dwarf goats have a low feed requirement.

Oberhasli

Oberhasli

Oberhasli goats were developed in the 1920s in the United States by breeding five imported Swiss goats to produce "chamois"-colored dairy goats. They resemble Alpine goats with bay or brown coats. They have two black stripes on their faces extending from their eyes to their muzzles, a black belly and legs, and a black forehead.

Oberhasli goats don't have as high a milk yield as other dairy goats, but the quality of their milk is excellent. Mid-sized goats, male Oberhaslis average about 150 pounds and stand about 30 inches tall. Females typically weigh about 120 pounds and reach about 28 inches tall. For small-scale hobby farmers, Oberhasli goats make a wonderful addition. They are trainable, mild-tempered, and curious.

Meat Goat Breeds

Raising goats for meat is a practice that is only recently becoming widespread in the United States. Culturally, the U.S. has preferred beef and pork to goat meat, but small-scale farmers are realizing that goat meat is lean, tasty, and nutritious. Adding meat goats to your hobby farm gives you control over the quality of the meat your family consumes, as well as control over the production and processing methods.

You can eat all breeds of goats, but some breeds are better suited for meat production. Here are some of the meat goats that are popular among hobby farmers:

Boer

Boer

The hands-down most popular meat goat for hobby farmers is the Boer goat. An import from South Africa, Boer goats are the variety that you will see at goat shows presented by 4-H, FFA, county fairs, and livestock shows. In the U.S., Boer goats have earned the nickname "Red Heads" because they all have white bodies and reddish-brown heads.

Boer goats are solid and muscular. The largest of all the meat goats, Boers are known for their fast growth rate, excellent meat quality, and good carcass yield. Full-grown Boer goats average between 200 and 300 pounds.

Additionally, Boer goats have a higher fertility rate than other goat breeds. Most goats have twins, but Boer goats commonly give birth to triplets and quadruplets. Small-scale hobby farmers should keep this in mind if they intend to breed their Boer goats. They could suddenly have twice as many newborn kids as they expected. The goat herd could quickly outgrow the accommodations at a hobby farm.

Kiko

Kiko

Kiko goats are meat goats that were first developed in New Zealand in the 1980s. Breeders Anne and Garrick Batten bred imported dairy goat bucks of various breeds with feral goats living in the area. The goal of the breeding program was to increase the rate of growth for the new goat breed, but the resulting goats, named Kiko after an indigenous word for meat, proved to be excellent meat goats.

A robust goat, the Kiko is well-suited for the wild and rugged pastoral environment of the New Zealand countryside. Kiko goats can be found in a range of colors and patterns. They have a comical bearded appearance. They have Roman profiles similar to Nubians and upright ears that point forward.

Hobby farmers raising meat goats are attracted to Kiko goats because the animals are naturally parasite-resistant. Female Kiko goats have an easy time kidding their young and exhibit a strong maternal instinct, meaning hobby farmers won't need to bottle feed the kids.

Savanna

A newcomer to the meat goat scene, the Savanna goat breed was first imported to the United States in the 1990s. A white goat, the Savanna can be confused with the Boer goat, and indeed, this breed was developed using Boer goat breeding stock. The result is a large, muscular meat goat that is hardy and adaptable.

Savanna

Unlike some other goat breeds, Savanna goats are not seasonal breeders. That means they can be bred any time during the year to spread out kidding time. They have a high fertility rate, and most births result in twins.

Savanna wethers, castrated males, have a particularly impressive growth rate. They mature early and produce good carcass yields.

Currently, goat breeders in the U.S. are continuing to work to further develop this breed. In fact, breeders are using artificial insemination techniques to prevent inbreeding problems.

Spanish

The small-sized Spanish goat breed originated with goats that were brought to Texas in the mid-1500s by Spanish explorers. A stocky, robust animal, the Spanish goat produces tasty, quality meat, but not as much as larger-breed meat goats. This makes them ideal for hobby farmers who want to raise meat goats for their own consumption, not for commercial production.

Spanish

A drawback of the Spanish goat is that the breed does not have a consistent rate of growth and productivity like other goat breeds. Additionally, the breed has been described as "flighty." These goats are active, curious, and smart enough to escape their enclosures.

The Spanish goat has an unusual nickname, the brush goat. That's because these goats are such good foragers that they are often used by farmers and ranchers to clear out brush and overgrown areas—a necessary method to reduce the spread of wildfires.

Spanish goats have a diverse range of fur colors and coat patterns. Since the breed was commonly bred with Nubians, Spanish goats have the same pendulous ears that we see on Nubian goats.

Dual-Purpose Goat Breeds

For small-scale hobby farmers, it is important to maximize space and get the most output possible from a small goat herd. For this reason, many hobby farmers look to dual-purpose goats as a way to have the best of both worlds—dairy and meat.

As you might suspect, dual-purpose goats do not excel in both categories, meat and milk output. They will, however, give hobby farmers adequate performance in both areas. Here are a few of the more common dual-purpose goats:

Sable

Sable

Sable goats, a Swiss breed, are nearly identical to Saanen goats, with one notable exception. While all Saanen goats are white, Sable goats can be any color or combination of colors and patterns, with the exception of white. It all comes down to the genes. The gene for white coloration is dominant. If the animal has this gene and is white, it is a Saanen goat. If it lacks this gene and is any color other than white, it is a Sable goat.

Sable goats, as well as Saanen, are classified as dairy goats because of their relatively high milk output. They are large goats with good muscle tone and a stocky physique, which is a sought-after trait in meat goats as well.

Kinder

Kinder

The Kinder goat breed got its start on a farm in Washington State in the mid-1980s. Breeders there crossed Nubians with Pygmy goats and selectively bred the resulting stock to come up with the Kinder goat. A muscular yet compact breed, Kinder goats are well suited for use as dairy producers or for meat production.

On average, Kinder goats gain weight quickly and generally have a dressed weight of about 60% after slaughter. As for milk production, Kinders have an average milk output of about 1,500 pounds per year and a lactation cycle of about 305 days. The milk is quite high in milk solids, making it ideal for making cheese.

Kinder goats are medium-sized. They get their stocky frames from their Pygmy goat ancestors and their long, slender legs from their Nubian heritage. Does typically stand between 21 and 26 inches tall at the shoulders and weigh about 115 pounds. Bucks can reach up to 28 inches tall and about 135 pounds.

Pygmy

Pygmy

The American Pygmy goat, which is usually just referred to as Pygmy, is a dwarf variety, just like the Nigerian Dwarf goat. Pygmy goats were developed between 1930 and 1960 by breeders in the U.S. Initially, Pygmy goats were bred for use as research animals or for petting zoos, but folks were fond of the diminutive goats that were full of personality. Beginning in the 1970s, Pygmy goats were bred more as companion animals than for scientific research.

The compact and stocky Pygmy goat stands between 16 and 22 inches tall at the shoulder and has a weight range of 55 to 90 pounds. The breed standards for the Pygmy goat state that seven color and pattern variations are acceptable: caramel with brown, caramel with black, black with white, solid brown, solid gray, solid black, and black banded.

Today, hobby farmers often raise Pygmy goats for meat. As milk producers, Pygmy goats do not have a high milk output, but their milk is high in butterfat. They are also kept as pets or for petting zoos.

Specialty Breeds

There is more to goats than just meat and milk. As we mentioned with a few previous breeds, some goats have been developed to fulfill other roles, such as clearing brush or being in a petting zoo or laboratory animals. Specialty goat breeds have a unique appeal for hobby farmers because they bring something different to the table. Here are a few popular specialty goat breeds:

Fainting Goats

Myotonic, or fainting goats, can be used for meat goats, but most hobby farmers raise them for their entertainment value. Fainting goats have a condition called myotonia, which causes their muscles to freeze

Myotonic (Fainting)

temporarily when they are startled or spooked. Just do a quick search on YouTube or Google, and you will find plenty of videos of fainting goats doing what their name suggests—fainting. Most videos show the goats happily running around their pasture and then suddenly falling over. A moment or two later, they will be back up and running again as if nothing happened. This medical quirk has made fainting goats popular attractions on hobby farms and petty zoos.

Angora

Angora

Angora goats, a breed of goat from Turkey, have long, fibrous hair called mohair that grows as an undercoat to the goat's coarse outer coat. The mohair, which hangs like wispy ringlets, grows longer than the outer fur. The fibers, although not as fair and fine as cashmere, are woven into thick, warm fabric for blankets, sweaters, and coats. To obtain the mohair, the Angora goats must be shaved. The animals are typically shaved twice a year, giving the fur plenty of time to grow back.

Angora goats are petite and delicate goats. They stand less than 20 inches tall at the shoulders and have light, graceful bodies. Their small size, combined with the mohair, makes Angora goats a go-to favorite for hobby farmers interested in making their own textiles.

Cashmere

Cashmere

Like the Angora goat breed, the Cashmere goat's primary use is for textiles. This goat breed produces fine, downy hair on its winter under-coat. There are separate follicles from which these soft hairs grow, and the denser outer fur protects them. People have been using the soft, luxurious hair to weave into fabric for thousands of years. Cashmere wool sweaters and scarves are still considered to be quality, fashion-able, and posh.

Instead of shaving the Cashmere goats, the fine hairs are harvested by combing each goat. It is a labor-intensive process, but for hobby farm-ers, it is well worth the effort. Hobby farmers can control the quality of the wool in order to produce rare, exquisite cashmere items.

Do You Want a Buck?

When making your decision about which breed of goat you want for your hobby farm, there is another decision you need to consider. Do you want to keep a goat buck with your small-scale herd? As we mentioned in Chapter 1, a buck is an intact, mature male goat. A wether, on the other hand, is a castrated male goat. Keeping a goat buck on your hobby farm can have both advantages and disadvantages. Here are some points to ponder:

Boer Goat Buck

Pros

Breeding Stock

A buck provides the opportunity to breed your does and expand your goat herd. If you plan to breed your goats for milk, meat, or fiber production, having your own buck on your farm allows you to have greater control over the breeding process so that you get kids with the traits you most desire.

Controlling the Bloodline

Many goat breeders have a goal of developing their own bloodline. The only way to do this is to keep a buck or two. By owning a buck, you have more control over the genetics and bloodlines within your herd. You can select specific bucks to breed with your does, focusing on traits such as temperament, milk production, conformation, and appearance.

Freedom

You will eliminate the need to rely on external breeding services if you keep a buck in your herd. This gives you more freedom and independence since you can breed your goats on your own schedule.

Farm Aesthetics

Bucks often have impressive physical features, such as a robust stance and powerful horns. The presence of a buck on your hobby farm can add to the charm and aesthetic value of your farm. If your goal is to open your hobby farm to visitors, the buck may give an authentic appeal to your farm setting.

Cons

Smell

One of the biggest drawbacks to keeping a buck is the strong odor they emit, especially during rut or breeding season. The buck's musky scent can be quite pungent and may permeate the surrounding areas, which can be unpleasant for many people. This smell can be reduced by keeping the buck separate from the does and other animals and by regular, thorough cleanings of the buck's housing area.

Noise

Bucks are typically more vocal than does. This is especially true during breeding season when bucks use their vocalizations to attract does. Your neighbors may not appreciate the loud and persistent bleating of the buck. If you have neighbors living in close proximity to your hobby farm, you should consider the potential impact when deciding to keep a buck.

Aggression

Bucks can be territorial and will aggressively defend their space, particularly during breeding season. They may exhibit challenging behaviors such as head-butting, mounting, and causing injury to other goats and to humans. Proper handling, socialization, and management are needed when keeping a buck. You will need a separate living space and even more secure fencing.

Facilities

If you are considering keeping a buck, you need to make sure you have sufficient additional space and adequate facilities. Bucks need secure fencing to prevent them from escaping or hurting themselves or others. They also need to be housed away from does to prevent unplanned breeding and so you can control the breeding. Before you add a buck to your small herd, be sure you have the infrastructure in place to handle this new addition.

Food and Care

Bucks have specific care requirements. They eat more than does and are more aggressive grazers. They require routine health check-ups with special attention given to their reproductive health. You will need to maintain a breeding schedule and keep records of breeding.

Keeping a buck as part of your hobby farm changes the dynamic of your small goat herd. There are benefits to owning a male goat; however, there are challenges as well. Ultimately, the decision to keep a buck on your hobby farm depends on your goals, resources, and ability to overcome the obstacles associated with bucks.

Ways to Acquire Goats

After narrowing down your breed selection to the goat breed that will be best suited for your hobby farm, you then need to find some goats to purchase. Since there is no goat store at the local mall, how do you set about locating goats? It really isn't as difficult as you may think unless you are interested in an uncommon breed.

As with any purchase you make, be sure to do your homework, ask questions, shop around, and don't rush to buy the first goat that you see. Also, be wary of cheap or discounted goats—remember, you get what you pay for. Here are some suggestions to start your search for your new hobby-farm goat herd.

Local Breeder

Locate a local breeder in your area and inquire if they have goats for sale. Contact your county extension office, 4-H headquarters, or FFA chapter to see if they can point you in the direction of goat breeders in your area. You can attend livestock shows or agricultural fairs to contact breeders and inspect their animals. If all  else fails, try networking. Ask your friends, coworkers, and family members if they know any goat breeders they can recommend. Post your request on social media, such as Facebook, and you may be surprised at the responses you get. Networking within the farming community can often lead to finding quality goats from trusted sources.

Breeders' Websites

For many breeders, breeding and selling goats is a business, not a hobby; therefore, they run their farm like a business. They will have websites that offer information about their farm and experience. You will find color photographs of the goats, as well as pedigree information and awards any of the goats have won. They may include their breeding and goat-raising philosophy and management practices. There will also be contact information, such as an email address and telephone number, so you are able to reach out to them.

Online Marketplaces

Check online classified ad websites and marketplaces that are dedicated to livestock animals. You can find these on Facebook Marketplace, Craigslist, and other sites. Google the breed of goat you're looking for and your location, and let the internet give you a list of sources near you. Of course, you will need to carefully vet each of these sources to make sure they are reputable breeders who are committed to producing top-quality goats.

Goat Organizations and Breed Associations

Look for goat clubs, organizations, or associations in your region. Some groups are general and include all goat breeds, but others are breed specific. These groups can provide valuable resources and connections to breeders and individuals who have goats for sale. Many of them have online directories and forums where you can find information on available goats or get recommendations from experienced goat owners.

Livestock Auctions

Agricultural and livestock auctions are held in rural and farming regions across the country. Attend one near you and see what the goat availability is in your area. Livestock auctions also provide an opportunity to interact with sellers, inspect the goats, and potentially make a purchase. A word of warning about livestock auctions, however. Some goat farmers and breeders use auctions as a way to get rid of old, weak, sick, or troublesome goats. Thoroughly inspect the goat, ask questions about the animal's age, health, and condition, and seek advice from expert goat farmers before you make your purchase.

Feed Stores

Get to know the managers and key employees at the feed stores in your community. These people know the local farmers because they are the store's main customers. They might point you in the direction of local goat farmers, breeders, or hobby farmers who might be selling goats. At the very least, most feed stores have a bulletin board area that is a source of information that can be helpful. Business cards, flyers, ads, and more can be announced on these boards.

Summary

This chapter outlined many of the most popular goat breeds for hobby farmers who want goats for their milk, meat, or fibers. By under-standing the traits of each breed, you can select the goat breed that fits your needs and your goals. We also discussed the pros and cons of keeping a buck goat on your hobby farm. Lastly, we explained some of the ways that you can find breeders or goats for sale near you.

Photo Courtesy of Eva Jordan

CHAPTER 3

Setting Up Your Goat Farm

Well before you welcome your goats to their new home at your hobby farm, you need to make sure that your infrastructure is in place to properly house them. Goats aren't as delicate as horses, but they still need to have a good, solid shelter to protect them from the elements and tall, escape-proof fencing around their pasture.

A goat barn does not need to be elaborate or fancy, but it will make your life easier if it has a well-thought-out design, is large enough to accommodate your small goat herd, and has a few amenities. The goat barn we currently have on our hobby farm is actually our second one. The first one was irreparably damaged in a storm, so we temporarily moved our goats to the neighbor's house (thank goodness for good neighbors!) while we built our second goat barn.

Although it was a mess at the time—and required time and money we hadn't planned on spending—I now look at this destruction of our first goat barn as a blessing in disguise. It gave us an opportunity to reflect back on our previous housing system and make changes to improve our farm setup. We sat down as a family to discuss what we liked and disliked about the old barn and create a wish list for our

Photo Courtesy of Katelynn Sharp

dream goat barns. When all was said and done, we now have a larger, more solidly constructed barn that has a dedicated feed room and hay storage loft—things we didn't have before.

Before you jump into setting up your own goat barn and pasture on your hobby farm, I would recommend visiting some other hobby farms and talking to small-scale goat farmers in your area. Take a look at how they have their goat farm set up and ask them to explain what works well and what doesn't. This will help you to visualize your own farm and how you can apply this knowledge to your situation.

In this chapter, we will go into detail about setting up your hobby goat farm. We will discuss the types of suitable goat barns and the ideal size, as well as give you tips on must-have amenities, bedding options, and how to winterize your goat barn. We will then move on to fencing requirements and materials for your goat pasture. Lastly, we will talk about feeding and watering systems for your small-scale goat herd.

Designing and Constructing Suitable Goat Housing

Protective housing is one of the most important first steps you will take in ensuring the well-being of your small-scale goat herd. You will need to consider how best you can create a shelter that will keep your goats safe and comfortable while also fitting the goat barn into the space you have available. There will be a number of decisions to make and points to consider, including the following:

Size Requirements

Goats are fairly low-maintenance animals that don't have complex housing needs, but that doesn't mean they should be crowded together into a tiny space. Just how big should your goat barn be? Well, that depends on how many goats you have and how much time they will be spending in their shelter.

Ideally, your goats will have access to a fenced-in pasture or natural area where they can spread out and get exercise. They will only come indoors to sleep or to escape bad weather. If this is the case, you will need to plan about 15 square feet per goat when determining the size of your barn. This will give them adequate sleeping space. However, if your goats will not have free access to an outdoor space and will be spending much of their time in their barn, you will need more space. Plan for at least 20 square feet per goat for sleeping space and an additional 30 square feet for exercise space.

That means if you have five goats that have free access to outdoor space, your barn should be 75 square feet in size at a minimum. For five goats that will be primarily housed within the barn, you should plan for a minimum of 250 square feet for sleeping and exercise space.

Plan ahead when building your goat barn so you are able to add more goats to your herd if you want. You can either build a larger barn than you currently need, or you can design your barn in such a way that you can expand it with a building addition.

Consider Local Regulations

The size of your goat barn may be limited by local regulations or ordinances. One of your first steps should be to contact your community, town, or county to find out what the current rules and laws are regarding hobby farms and goat herds, as well as the regulations about the size and placement of outbuildings. You always want to operate within the confines of the law and abide by the rules. Know the laws before you build your goat barn—it will save you grief and hassle later on.

Goat Barn Requirements

The shelter for your goats should be designed to provide protection for various weather conditions. That means the goat barn must have a waterproof roof that doesn't leak, insulation to keep out the wind and freezing temperatures, and ventilation so airflow can keep the barn cool in the summer heat. In our goat barn, we also had an issue with rain-water running off the roof and creating a big, muddy mess in front of the door. We installed a rain gutter to direct the rainwater away from our entry point.

Windows will provide proper ventilation in the barn to maintain fresh air and minimize the accumulation of ammonia and moisture. Two of the windows in our goat barn were floor models from my husband's work. They were outdated models that were going to be discarded before my husband rescued them and installed them in our goat barn. Like a typical house window, they both have an insulated glass window that slides up to reveal a screen. We put these windows on the north side of the barn—the part that gets the most wind and rain. On the opposite side, we built a large, screened opening with a hinged wooden flap on the exterior. We can lift this flap and lock it open to allow air to flow through, then secure the flap tightly in place during the winter months.

Goats are susceptible to foot problems, so the flooring of your goat barn should be solid, dry, and nonslip. Packed dirt floors are fine as long as the dirt doesn't get too wet, muddy, and slippery. Concrete floors work well, too, but they cost more money, and the water doesn't drain away. Tile floors may be easier to clean, but they can be too slippery when wet. Your goats could fall and injure themselves.

Good-To-Have Features

After having a small goat barn for a few years and then being pre-sented with the opportunity to rebuild it to suit our own needs, we were able to include many features that are not really necessary but are cer-tainly convenient to have. One of these features is a feed room. Before, we stored feed in the garage and hay under a tarp. This was not an ideal setup. Our feed room has an interior wall that is made with slats of wood so air can circulate through. It also has a door that closes tightly and latches so an escaped goat with an appetite for destruction can't raid the feed room (believe me, they have tried!). There is a small loft above the feed room for hay storage, and in the feed room, there are bins for grain and shelves for various goat-related items. There is a door that opens into the manger from the feed room so we can fill the manger with hay more easily.

We also added a couple of stalls. Most of the time, the stall gates are left open so the goats can wander in and out as they please, but it is handy to have a place to keep newborn goats, a laboring doe, or a sick or injured goat where it won't be bothered by the other goats.

Our first barn had electricity, and we definitely wanted to keep that feature. But we also knew how we wanted to change it. In the old barn, the goats had access to the light switch. Like our teenagers, the goats left the light on all day and night with no regard for our electric bill!

There was also an outlet in the goat stall area, and we caught the goats nibbling on the outlet cover. I was afraid they were going to expose some wires and either zap themselves or start a fire. We ended up covering that outlet, and we made sure the outlets in the new barn were well away from the curious goats. We positioned a few outlets high up

on the walls so we could plug in electric fans and hang them out of reach of the goats. It helps to keep the temperatures down on the hottest days of summer.

We also added an exterior light outside the "people" door of the new barn. The switch to turn on that light is in the house. Now, we don't have to stumble out in the dark.

Another nice-to-have feature for a goat barn would be a water spigot and hose. This is one thing I wish we would have added to our goat barn. Instead, we still use the water spigot on the side of the house and haul buckets of water to the goat barn. It is not a fun job, especially in the wintertime. We have tried connecting several hoses together and stretching them from the house to the goat barn, but this has been problematic. The hose goes across the driveway, and we were constantly replacing hoses because someone drove over the metal couplings, flattening them. Oops.

When setting up your goat farm area, think about how you plan to clean the barn and remove soiled bedding material. If you plan ahead for waste management and cleaning, you can establish an efficient system to keep the barn clean. I already touched on flooring, but it is so important that I'm going to talk about it again. Whether you have a dirt or cement floor in your goat barn, try to slope it at enough of an angle that spilled water and animal waste drains to one spot. This will make clean-up easier and keep the majority of the barn dry for the goats.

Our barn floor slopes toward the barn door—the one the goats use to go to their pasture. In hindsight, that wasn't the best setup. While it is convenient to park a wheelbarrow at the goat door and shovel everything into it, the goats have to walk through the gross stuff on their way out the door. Our barn floor is dirt, so it is easy to rake it out. A cement floor can be kept clean by hosing it out.

Another option is installing a rubber floor. Rubber flooring comes in rolls with adhesive on one side so it can be affixed to a wooden or concrete floor. Like cement floors, rubber floors are easy to clean because they can be hosed down and mopped. The rubber is softer than concrete, so it is more comfortable for the goats. A drawback of rubber flooring is that it can stay wet longer than other flooring materials. Goats are susceptible to hoof rot if their flooring is too wet for too long.

A few other suggestions regarding the floor. Add rubber baseboards to protect the walls of your barn from the almost constant moisture. And keep the stall space simple. The more corners, nooks, and crannies, the harder it will be to clean.

How do you plan to dispose of the soiled bedding from the goat barn? Many hobby farmers opt to compost the bedding, which is a great idea. "Nanny berries," a colloquial term for the small, round, pellet-like goat poop, are high in nitrogen and make excellent fertilizer. We set up our compost pile a short distance away from the goat barn in an out-of-the-way place. The soiled bedding gets dumped there.

Building Materials

Most barns are made out of wood, although I have known some goat farmers who have converted metal storage sheds and grain bins into goat barns. A metal goat barn, I would think, might be very hot in the summer and cold in the winter, but if you have access to a free or low-cost metal shelter for your goats, I'm sure you can make it work.

We built both our goat barns using wood—my husband works in the lumber/building materials industry, so he has access to discounted building supplies. He even gets free items that are being discarded.

In our first barn, we used particle board, the stuff that is made with compressed wood chips held together with resin or glue. Our weird goats loved the taste of the glue and began licking the walls of the goat barn. Licking turned into nibbling. Pretty soon, they chewed holes in the walls. We patched the holes with more particle board, but they still destroyed it. I think they were able to loosen some of the wood chips enough that they could get their teeth around them. We had to go with plywood instead, and they now leave the walls alone.

Keep this in mind when you are selecting building materials for your goat barn. You don't know if your goats will be barn-nibblers like mine, and you don't want them to ingest materials that will be toxic to them. Some treated lumber, wood stains, and paints can be toxic to goats. Look for products that are "livestock safe."

Bedding

Bedding material placed in the goat stall area of your barn will provide a comfortable resting place for your goats. Bedding also absorbs urine, nanny berries, and other moisture. Lastly, bedding provides insulation in cold weather that will help keep your goats warm.

You can use a variety of different bedding materials, depending on your preference. Straw, wood shavings, shredded corn cobs, and

*Photo Courtesy of
Mary Beth O'Donovan*

shredded newspaper are all options. I have even read that you can use the leaves you rake from your yard in the fall as goat bedding. I tried this once. Instead of sleeping on the bedding, the goats ate the leaves.

As the weather gets colder, add more bedding materials to the goat barn to act as an insulator against the cold. Be sure to regularly remove and replace the goat bedding to maintain the cleanliness of the barn and to prevent the buildup of odors, moisture, and parasites.

Winterization

If your hobby farm is located in a region that experiences brutal winters, keep this in mind when you are setting up your goat barn. There are things you can do to reduce the impact of cold temperatures on your herd. Insulate the walls of your goat barn and seal up cracks and holes where the cold can creep in.

Think about the placement of the doors and windows of your goat barn. In our original barn, the goat door faced to the northwest, the

direction the strong, brutally cold winds came from. When we opened the goat door to allow the goats to roam out into the pasture, the wind, snow, and rain blew right into the barn. When we rebuilt, we moved the door to the other side of the barn. That way, the opening was protected from the elements.

All doors and windows should be solid and able to close securely. In the winter,

DID YOU KNOW
World's Largest Goat

The Guinness World Record for the world's largest goat belongs to a British Saanan owned by Pat Robinson and Ewyas Harold in the United Kingdom. This goat, named Mostyn Moorcock, grew to a withers or shoulder height of 44 inches and a length of 66 inches. British Saanan have slightly longer legs than Saanan goats and can produce an average of 11 pounds of milk daily.

latch them shut or lock them. If the wind can still blow in through the windows, consider installing winterizing plastic over the windows—on the outside of the barn, where the goats can't nibble it.

You probably don't need to use a heater in your goat barn to keep your goats warm in the winter. They are hardy animals, and they give off enough body heat to keep each other warm. Besides, it is extremely dangerous to use space heaters in general—but using them in a place that is packed with flammable straw and hay seems like a really bad idea.

Remember that polar vortex winter we had a few years ago? We were naturally concerned about our goats during that unusually frigid time, but my daughter was particularly worried. She rummaged through her closet and found a few old sweatshirts she could bear to part with. And, yes, she put the sweatshirts on the goats! They looked ridiculous—but it kept them warm.

Creating a Secure Pasture

A goat barn is only half the story. You will also need a secure pasture to give your small-scale goat herd a safe place to run around, get exercise, and forage for snacks. Your pasture should be roomy enough to accommodate your goat herd, secure enough to keep predators out, and tall enough to keep your goats from escaping.

Ideal Pasture Size

When you talk to other hobby farmers, you will get a lot of conflicting information about pasture requirements. Some will tell you that their herd is perfectly content in a small, fenced-in run, while others will insist you need acres of pastureland. You need to find a happy medium.

At the absolute minimum, you should provide about 200 square feet of pasture space per goat. Since you will have at least two goats—goats are pack animals that need companionship, and they get depressed living alone—your pasture should be roughly 400 square feet. But remember, that is the bare minimum. Ideally, your goat herd should have more space to roam and nibble.

Goats, as natural browsers and foragers, instinctively want to look for plants and leaves to nibble on. They will eat down all the vegetation in a small pasture in no time and will be anxious when they can no longer find food. Besides, if they eat all the grass and weeds, you will need to provide them with more hay and grain to make up the difference.

My advice regarding pasture space would be to go big. Your goats will thrive with plenty of room to exercise, and your hobby farm will be ready to accommodate more goats if you decide to expand.

Keep Out Predators

Fortunately, goats are large enough that most predators, like raccoons, foxes, and hawks, will leave them alone, although there have been cases of young kids being attacked by these animals. Most likely, the biggest threats to your hobby farm goat herd will be coyotes and domestic dogs. Your pasture fencing should aim to keep these predators out.

I live in an area with a high coyote population. We have lost some chickens to coyotes, but they have never bothered our goats. That could be because the chickens were easier to catch and posed less of a threat to

the coyotes. Coyotes are opportunistic hunters who are big into self-pres-ervation. They don't want to get into a fight with something that might harm them. Coyotes, however, often hunt in packs. They certainly could take down a goat if they wanted to.

Your neighborhood dogs are really just the domesticated cousins of the coyotes. Even though pet dogs enjoy all the pampering of domesti-cation, deep down, they still have the natural instinct to hunt. A running animal is like an invitation to chase and attack dogs. Some breeds have a stronger chase instinct than others, but you should be aware that any dog has the potential to harm your goats.

We have a dog—a very large one. When he was just a playful puppy, he liked to chase the goats and nip at their feet. Brown Sugar put him in his place by doing what goats do best— head-butting. She butted our dog in the head, sending him somersaulting backward. It only took a few knocks to the head to instill in our dog a healthy dose of respect for the goats.

In my experience, a fence that is tall enough to keep a goat from jumping out will also be tall enough to keep a coyote or dog from jumping in. But that's not how dogs operate. Instead of leaping over a fence to get what they want, they will dig under it. You can take the added step of keeping your herd protected by burying the fence several inches into the ground. Dogs and coyotes will have a tough time burrowing under the fence to get to your goats.

Fence Heights

Goats are agile jumpers. Larger goat breeds, like Nubians or LaManchas, can easily clear a four-foot fence with a running start. They are also determined and clever animals. If there is something on the other side of the fence that has their attention—like my pear trees—they will work hard to reach their goal.

Most goat experts suggest that your pasture fence be at least five feet tall. I would go a step further and recommend a six-foot fence, so you don't have to deal with an Olympic high jumper like Dixie.

Escape Proof

Leaping over a fence is just one way that goats can escape from their pasture. There are a million other ways. Goats are creative problem solvers. They will look for any weakness in your fencing and exploit it. In all fairness to Dixie and our other Houdini-ish goats, we have never had a goat run away. When Dixie or one of her friends escaped the pasture, they just hung out in the yard, eating my pear saplings or the basil growing in a container on the porch, perching on top of the car, or yanking laundry off the clothesline. Thankfully, we never had to deal with an irate neighbor because one of our goats paid a visit to their yard. Instead, Dixie seemed to take delight in watching us pull into the driveway and immediately start chasing her. If goats had thumbs, I'm sure she would have thumbed her nose at us.

Pay particular attention to your gates. We have had goats that have figured out how to lift up the U-shaped latch and push the gate open.

*Photo Courtesy of
Katelynn Sharp*

Others learned to squeeze through the small space between the gate and the post.

Another weak spot is where one section of fencing ends and another begins. Goats can spot this from across the pasture. Somehow, they know that they can loosen the fencing in this spot by pushing hard against it and wiggling the fence post nails free.

If there is a loose fence pole, your goats will find it. If there is a hole in the fence, they will exploit it. If there is a gap anywhere, they will use it to their advantage. Goats should work in quality control. Like the character in that Eagles song, "Desperado," you will need to "ride the fences" or just walk around the pasture from time to time, looking for places that need repairs.

Fencing Materials

Pay a visit to your local farm and feed store or a home improvement store, and you will find that there are many different types of fencing on the market. What is the best kind to use? It might be easier to explain

what kinds of fencing don't work for goats first before we talk about viable options.

Chicken wire, with its hexagon-shaped openings, is charming to look at but impractical to use, especially for goats. The wire is flimsy and easily bent. A goat can easily crumple it during an escape attempt. Likewise, mesh fabric fences, like the kind people install to prevent the snow from drifting, are simply not sturdy enough to keep goats confined in their pasture space. Electric fences are generally frowned upon for goats. They can cause harm and injury to a goat. I heard a story once about a goat that got its horns tangled in an electric fence and was repeatedly shocked for a few hours until the farmer returned home to rescue the poor animal.

Chain-link fencing is a good option; however, it can be expensive. When you install it, you will need to make sure it is pulled as tightly as possible, or there will be sags and dips in the fencing. Goats will use this to plot their escape. Another good choice is galvanized woven wire fence panels, the kind with square holes, as this material is strong, durable, and will not rust. Look for a variety with smaller holes. Goats can get a foothold in larger holes and climb over the fence. Cattle panels will also keep goats penned up, but these are also quite costly and may not be tall enough to prevent larger goats from leaping over them.

Pasture Amenities

When planning where to position your pasture, look for natural features that will be beneficial for your goats. If at all possible, try to include tall, well-established trees in the pasture area. Goats need shade from the sun. If the trees are smaller, however, the goats will strip the branches and pull off the bark. We have a few large trees in our pasture that are so mature that the lowest branches are well out of reach for the goats. Of course, you will need to make sure that all the plants growing in your pasture area are safe for your goats to eat. In Chapter 7, we will go into detail about the trees and plants that are toxic to goats.

Goats like to climb, and it helps keep their hooves worn down to scramble over rocky terrain. If you have some large rocks on your hobby farm, it might be a good idea to include those in your goats' pasture.

We had an overgrown fieldstone foundation from some old outbuildings in the backyard. It was an unsightly mess. We put the pasture fencing around it, and in just weeks, the goats had eaten down the scrubby trees and weeds growing up around the foundation walls. Now, the goats have a few rocky walls and ledges they like to climb on.

Photo Courtesy of
Gayle Ewen

Establish a Feeding and Watering System

When setting up your hobby goat farm, plan how you will handle feeding, watering, and feed storage. We failed to do this when we built our first barn, which is why we had to store the feed in the garage and stack the hay against the side of the barn under a tarp. It is much easier, we discovered, if you make room for these things from the onset.

Water Containers

Your goats need to have constant access to fresh, clean water. There should be a waterer inside their goat barn and in their pasture. Here's one truth I have learned about goats—they like to knock over water buckets. It's their favorite thing to do. After cleaning up their water mess and replacing damp bedding for the hundredth time, we rigged up a way to keep the water bucket securely upright. We felt triumphant for a few moments until we realized that we would have to work to unstrap the bucket and restrap it every day when it was time to wash and refill it. What a pain! We tried a wider-based waterer, thinking that the goats wouldn't be able to tip it over, but we soon discovered that the goats are much smarter than we are. The bucket inside the barn is still strapped in place.

As for the water container in the pasture, we have a small, galvanized water trough that is heavy enough to stay in place. We placed it at the point in the pasture that is closest to the water spigot but not in a place where leaves and twigs from the trees can fall into it. There is a valve at the bottom to empty the water, and we stretch the garden hose over to fill it up. It is admittedly not the most efficient system, but it works.

Feed Containers

For grain, we use a long, narrow feed trough that fits into a wire frame attached to a wall in the goat barn. I believe we purchased it at our local farm and feed store. The plastic trough snaps out so we can easily wash it, and it snaps back into place so securely that the goats haven't yet figured out how to remove it.

The long trough system is ideal. Goats are bullies, especially when it comes to eating. They will push each other out of the way and eat all the food. By spreading out the food, all the goats can eat at once.

In addition to the feed trough, we have a hay manger in the goat barn. It looks like an oversized wire magazine rack and holds a large fold of hay. The goats can nibble on the hay all day. They are messy and wasteful eaters, so there is always a pile of uneaten hay on the floor under the manger.

Hay Storage

Hay becomes moldy and spoiled when it becomes wet; therefore, it should be stored in a dry place. Now, we have a small hay loft and feed room within the goat barn to store the hay. As a small-scale goat farmer, you will have trouble finding a hay farmer to sell you small amounts of hay. They prefer to sell to cattle and horse farmers who buy in bulk. You may have to buy more hay than what you really need, so storing it properly becomes important.

Winter Feeding and Watering

Freezing temperatures create a whole new set of problems for hobby farmers. It is a constant battle to keep the goat water from freezing. Fortunately, there are heated waterers on the market that will prevent the water from freezing. Grain-based goat feed has enough moisture in it that it can freeze together in the winter. You may need to temporarily move the grain into the house or garage to keep it from becoming a solid chunk.

Summary

Getting your hobby farm ready to bring in a goat herd is an exciting time. You may be tempted to rush through the preparations so you can welcome your new goats as soon as possible, but it is better to take your time and make sure the goat barn and pasture are well planned out and thoughtfully designed.

As we learned in this chapter, the size of both the goat barn and the pasture space is important for the health and well-being of your small-scale herd. You will need to ensure that the building materials you use are safe for livestock and that the floor of your goat barn is conducive to cleaning. You will have to consider the different types of bedding options available to you and decide on the best food and water containers. When building your goat barn, you should also include strategies to winterize the building to keep out the cold, rain, and snow.

The pasture space needs to be roomy, secure, and escape-proof; therefore, the fencing materials you select are key. The pasture should be constructed with the goal of keeping potential predators out and the goats in. Goats will uncover any weak spots in your fence and make their escape, especially if there is something tempting on the other side of the fence—like apples on a branch or tomatoes on a vine.

With proper planning, you will be able to create a safe haven for your goats and a functional area for your hobby farm.

CHAPTER 4

Basic Goat Health Care

Your small-scale goat herd depends on you to take proper care of them. Goats that are well-cared-for and have an attentive hobby goat farmer to see to their basic upkeep will thrive and be less likely to experience medical problems. Goats are, for the most part, low-maintenance animals. They don't require a lot of time-consuming grooming or fussing on a daily basis; however, you will need to do some routine goat care periodically, which we will outline in this chapter.

Goats need access to fresh water and nutritious food every day. In Chapter 7, we will go into detail about the nutritional requirements of goats. Chapter 5 will discuss common diseases that impact goat health, and Chapter 12 will explain how to handle an emergency health or medical situation involving your goats. I invite you to enjoy the humorous anecdote about the time one of our goats fell down an old well shaft. (Don't worry—it has a happy ending. An embarrassing yet happy ending.)

To understand the needs of your goats, you should have a working knowledge of basic goat anatomy. It can be helpful to have this information so you know how you can best meet their needs. My kids spent many years in 4-H showing goats at our county fair. Part of the showmanship competition at the goat show involved memorizing goat anatomy and functions. Since my kids were super-dedicated to summertime studying (insert eye roll), I spent hours the night before the goat show quizzing my kids about goat parts. As a result, I am quite knowledgeable about the anatomy of a goat. My children, however, are not.

Basic Goat Anatomy

Head and Neck

The goat's muzzle encompasses its mouth, nostrils, snout, chin, and lips. Many goat breeds have beards growing from their chins. Female goats can have beards like male goats. Goats' upper lips are prehensile, meaning they can grasp and hold leaves and branches, even pulling the foliage from trees and bushes. Instead of upper incisor teeth, goats have dental pads on their upper jaws. The incisors and premolars along their lower jaws are the teeth they use for chewing. The neck area doesn't just cover the front or throat area. It also includes the back of the animal's head and extends to the shoulder area to the withers. Goats' eyes can look rather creepy at first. That's because their pupils are long, horizontal

rectangles, not the round circles that humans have. This adaptation helps goats keep an eye out for predators, even when they are grazing. A number of goat breeds naturally have horns while others are polled or hornless. Female goats can also have horns, so the presence of horns should not be an indicator of gender. The shape of a goat's ears is determined by its breed. Some are large and pendulous, others are perky and upright, and still others seem to be nonexistent.

Legs and Hooves

Goats have long, slender legs that are strong and agile, perfect for climbing. The elbows of the goat are not where you would assume. The elbows are the highest joint on the goat's front legs and are located just below the animal's chest. What you would consider the elbows are actually the goat's knees. The portion of the front legs between the elbows

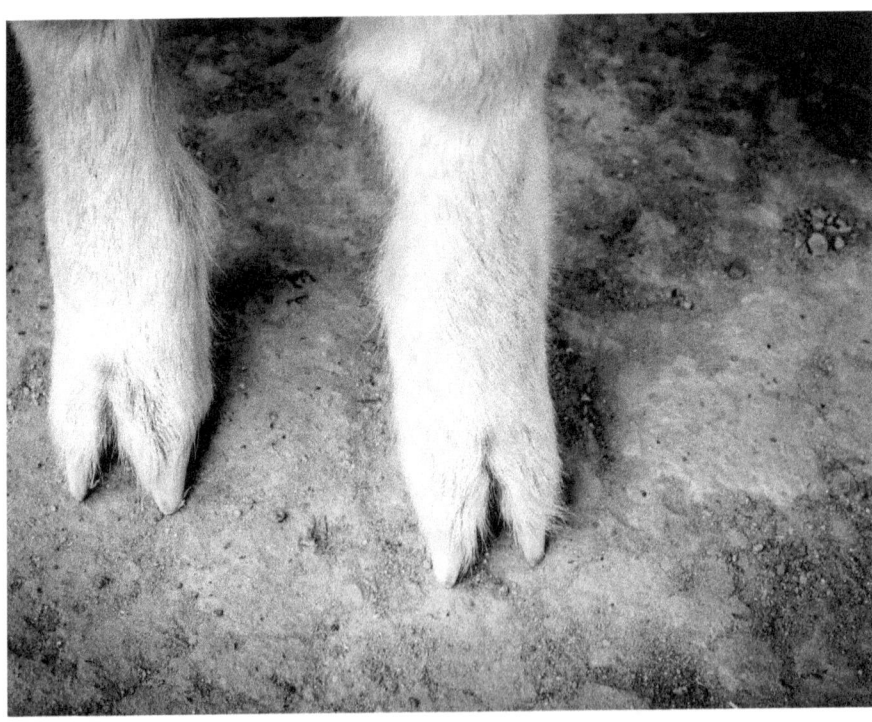

and the knees is called the foreleg. The thighs are the highest part of the goat's hind legs. The forward-facing, rounded, bony part of the goat's hind legs is called the stifle, while the rear-facing, pointy, bony part halfway up the animal's hind legs is the hock. Goats have cloven hooves, meaning they are divided into two distinct toes. Behind the front hooves, a goat may have dewclaws—small, fleshy appendages. The pliable area between the dewclaw and the hoof is known as the pastern.

Body

The entire area between a goat's neck and its legs is called the back. The highest portion of the goat's back is its withers. From the withers, the animal's shoulder is the thick bone that extends downward to the

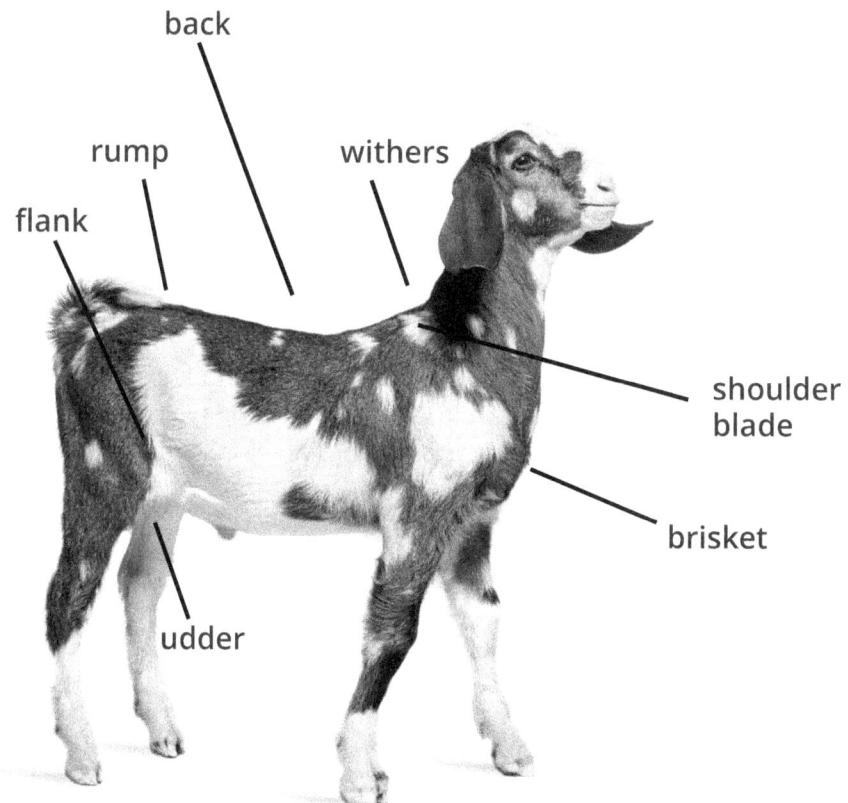

back

rump

withers

flank

shoulder blade

brisket

udder

chest. The top back of the goat is its rump, and that's where you will find the animal's tail. On either side of the animal, just beyond its head, are the shoulders and brisket. The same region on the back end of the goat, below its rump, is the flank. A female goat will have udders on her underside, between her hind legs.

Digestive System

Although you can't see a goat's digestive system, you should understand how their digestive system differs from humans. The digestive system of goats evolved to help them process plant material more efficiently. As members of the rudiment family, goats have four-chambered stomachs—the rumen, reticulum, omasum, and abomasum. Fresh green leaves, woody branches, and dry hay pass through each chamber in a complex process that breaks down the cellulose in the plant material to extract the nutrients.

Familiarize yourself with the various parts of the goat's anatomy and observe how they look each day. A healthy goat will have bright, alert eyes, a shiny coat, and clear, clean nostrils. You will be able to quickly see if there has been some change that could indicate an illness or injury. You should also take note of typical goat behavior. A thriving goat will be active, curious, and playful. On the other hand, a goat that is feeling under the weather will lack energy, seem disinterested, and may stay in one spot for most of the day. Take the time to get to know your small-scale goat herd. Once you learn the personality and typical behavior of each goat, you will be more in tune with problems, should they arise.

Preventative Measures

Prevention is the best medicine for goats. Routine care and preventative measures will help keep illness and injuries at bay.

Vaccinations

Vaccinations are intended to prevent diseases, particularly diseases that can spread to other goats in your herd. Many experienced small-scale goat farmers vaccinate their herd themselves; the vaccines are sold online and at farm and feed stores. I've vaccinated my goats. I asked a longtime goat farmer in our 4-H club to give me a lesson on injecting goats, and I've done it myself ever since. A word of warning, though—just because you have easy access to vaccines doesn't mean you should over-vaccinate your herd. Most goat experts take a minimal-ist approach to vaccinations. After all, these are substances that you are injecting into your goats' bodies and could impact the milk or meat they produce. On the other hand, some vaccines are necessary to ensure the continued health of your herd. Discuss vaccines with your trusted veterinarian and follow their guidance. The majority of veterinarians suggest the vaccines listed below as part of your goats' routine care. Depending on where you live and other factors, your veterinarian might recommend additional vaccines.

Vaccine	Protection Against...	When to Administer
CDT	Enterotoxemia and Tetanus	Kids: At 1 month old and 2 month old Does: During the 4th month of pregnancy All: Annual booster
Pneumonia	Pneumonia multo-cida/Mannheimia Haemolytica Pneumonia	Kids: At 2 month old repeated 2 to 4 weeks later
Rabies	Rabies	Annually
Soremouth	Orf	Annually
CLA	Corynebacterium pseudotuberculosis	Kids: At 3 month old repeated 3 weeks later All: Annual booster

Deworming

Goats are susceptible to internal parasites and can get worms in their lungs, stomachs, and livers. Parasites can damage the animal's organ tissue and cause anemia. The solution is to give your herd a dewormer; however, this is a matter of debate among goat farmers. In the past, the conventional method was to administer dewormer on a routine schedule. Unfortunately, the overuse of dewormers has created strains of parasites that are resistant to the medication. Now, many goat experts recommend deworming your goats on an as-needed basis, not as a routine preventative. They suggest frequently monitoring your goats for signs of worms or anemia and only treating known outbreaks of parasites. I live in an area with hot, humid summers—ideal conditions for parasite outbreaks—so I deworm them once a month in the summer. I sure hope I am not contributing to the problem.

In the past, goat kids were given their first dose of dewormer when they were about four to six weeks old, then once a month for the first four or five months. After that, they were put on the same schedule as adult goats—deworming every four to six weeks. Today, it is recommended that you inspect your goats for signs of anemia by looking at their gums and under their eyelids. The tissue in these places should be red or bright pink. If they become pale pink or white, it is an indicator of anemia caused by a parasitic infection that requires treatment with a dewormer.

Goat deworming products are widely available online and at farm and feed stores. Most often, they come in a paste form in a tube. The correct dosage is measured using a device that looks like a caulk gun. Then, you just have to hold the goat still and squirt the stuff in its mouth. It has been my experience that goats find the deworming medication to be quite tasty. They eagerly lap it up.

Many hobby farmers who are new to goat raising mistakenly believe that internal worms happen because parasite eggs hatch inside the goat's body. That's not how it works. Instead, the worm eggs hatch on the grass of the pasture. When the goats eat the grass, they ingest the larvae too. I am telling you this to explain that you can greatly reduce the likelihood of your herd consuming too many of the newly hatched worm larvae, leading to an infestation of parasitic worms. How do you do this? By rotating your pastures. Move your herd from pasture to pasture often so they don't spend too much time in one area.

You can try to keep parasites down by using natural food items with deworming properties. Many experienced goat farmers swear by pumpkin seeds. Science backs this up—to a point. There are properties in pumpkin seeds that are naturally antiparasitic in nature. The quantity of these properties, however, is fairly low. Your goats would have to consume large amounts of pumpkin seeds for it to be totally effective. But it certainly doesn't hurt to add pumpkin seeds to your herd's list of treats. My goats love them, but they don't get them very often. When we carve our Halloween jack-o-lanterns each year (for the super-competitive family pumpkin carving contest. Seriously, we go all out!), the goats enjoy a nice bowl of pumpkin guts.

Pine needles, which contain tannins, can also act as a natural dewormer. As with pumpkin seeds, pine needles won't completely solve the worm issue, but they can help.

Dehorning

Some goat breeds are naturally hornless. Others are not. Kids are born without horns, and the horns grow as they mature. Horns are not a desirable feature in goats, especially hobby farm goats. Without horns, goats can be housed closer together. For hobby farmers with limited space, this is ideal. The horns can injure other goats, small children, dogs, adults, and even the goats themselves. If you plan to welcome visitors to your hobby farm, your guests will be safer around dehorned goats. A goat's horns can also become tangled in fencing, tree trunks, or other places.

The process of dehorning goats, however, is not pleasant. It can cause pain and stress on the goat. There is also the risk of infection, excessive bleeding, and scurs, which are bony spikes of horn growth from incomplete dehorning.

Dehorning, also called disbudding, is usually done when the goat kids are only a few weeks old. I'll be honest—it is a disturbing process. The baby goat is restrained, sometimes in a wooden box with an opening for its head. The hair around the tiny horn buds is shaved or trimmed to fully

expose the horn buds. Sometimes, a local anesthetic is applied. Then, the person performing the procedure presses a hot debudding iron or cauterizing tool on the horn buds to burn them, destroying them and preventing further horn growth. It only takes about a minute per horn to cauterize it, but, unfortunately, it seems extremely painful for the kids.

I would highly recommend that you have a veterinarian do the debudding or find an experienced, qualified person who can do it for you. I have not dehorned my goats myself. It is such a brutal-looking procedure that I don't have the heart to do it myself. I have, however, witnessed it being done, and it was tough to watch. However, one of the reasons we got goats was for my children to participate in 4-H goat shows at our county fair. Only dehorned goats were allowed, so debudding was necessary. Fortunately, the baby goats quickly recovered from the procedure and went happily bouncing around the pasture. We never had any goats develop infections or complications from the procedure.

It is possible to remove the fully developed horns from an adult goat, but most goat experts strongly discourage this. It is a bloody, brutal, and traumatizing procedure for the goat. The goat is restrained to ensure the safety of the animal and the person performing the procedure. A local anesthetic or sedative is used, under the supervision of a veterinarian, to keep the animal calm and to reduce stress. The horns are sawed off as close to the animal's skull as possible. Then, the entire area is cauterized to stop the bleeding and prevent the growth of horn scurs. The wound site on an adult goat is much larger and more invasive than that of a kid; therefore, aftercare is particularly important. There is an increased risk of infection, bleeding, and further complications. The potential for infection is so high, in fact, that most goat experts and veterinarians advise against dehorning adult goats.

Castrating

There are many reasons why male goats are routinely castrated. The obvious reason, of course, is to prevent unwanted goat pregnancies. This is especially important for small-scale hobby farmers who don't have the space or resources to handle an uncontrolled population boom in their herds. But there is also the issue of smell. An intact male goat, or buck, will spray his scent all around the area as a way to mark his territory and claim his does. It is a nasty odor that will make it unpleasant to be in your goat barn. The neighbors might even complain. Lastly, castrated male goats, or wethers, produce better-tasting, more tender meat.

Like debudding, castration should be done when a goat is still a kid. The ideal age to castrate a kid is between 8 and 12 weeks old. As with debudding, this procedure can be done by a veterinarian or by an experienced goat farmer who knows what he or she is doing. For the record, that's not me. I have not castrated goats myself, but I have been present when it was done. There are a few different ways to castrate a goat.

Surgical castration is performed by a veterinarian in a sterile operating room. The veterinarian will either sedate the goat or administer a local anesthetic to minimize pain. A small incision is made at the bottom of the scrotum and the testes are completely removed. The wound is closed with sutures, surgical glue, or staples. Surgical castration results in a complete halt of testosterone production in the goat.

The clamping method, also known as the Burdizzo method, requires the use of a special tool called a Burdizzo. The goat is restrained and either sedated or given local anesthesia to dull the pain. The scrotum area is cleaned and disinfected before the clamp is positioned above the goat's scrotum. The clamp is then squeezed to crush the blood vessels and spermatic cords entering the animal's scrotum. The clamp is held in place for a designated time to make sure that the vessels and cords are fully crushed. The clamping method is relatively quick, but the biggest benefit is that it is bloodless. The risk of infection is much lower because there are no open wounds.

The banding method of castration involves placing a strong rubber band or elastic ring tightly and securely around the scrotum to cut off the blood flow to the testes. The area is first thoroughly cleaned and disinfected. The kid is then secured, and its scrotum is exposed. The band is placed above the testicles and tightened. The lack of blood flow to the testes causes them to atrophy and fall off. The banding process is simple, cost-effective, and bloodless; however, it can take several weeks for the testes to fully die and detach. There is a risk of the area becoming infected.

Chemical castration requires the use of hormone inhibitors to suppress testosterone production. The hormone inhibitors can either be injected into the goat's body or a slow-release implant can be placed under the animal's skin. This method is more costly than some of the others. If the dosage of the hormone is not correct, the animal may not be fully or permanently castrated. It may require repeat treatments to get the job done.

Hoof Trimming

Goats lived in the wild for thousands of years with no one around to trim their hooves, so why do you, as a hobby goat farmer, need to worry about hoof trimming with your small-scale herd? It all has to do with a side effect of domestication. Goats in the wild spent their days climbing on rocky slopes, over boulder-strewn hills, and over rugged terrain. Their hooves, which naturally grow continuously, were constantly being worn down by their environment. Goats living on a hobby farm don't have

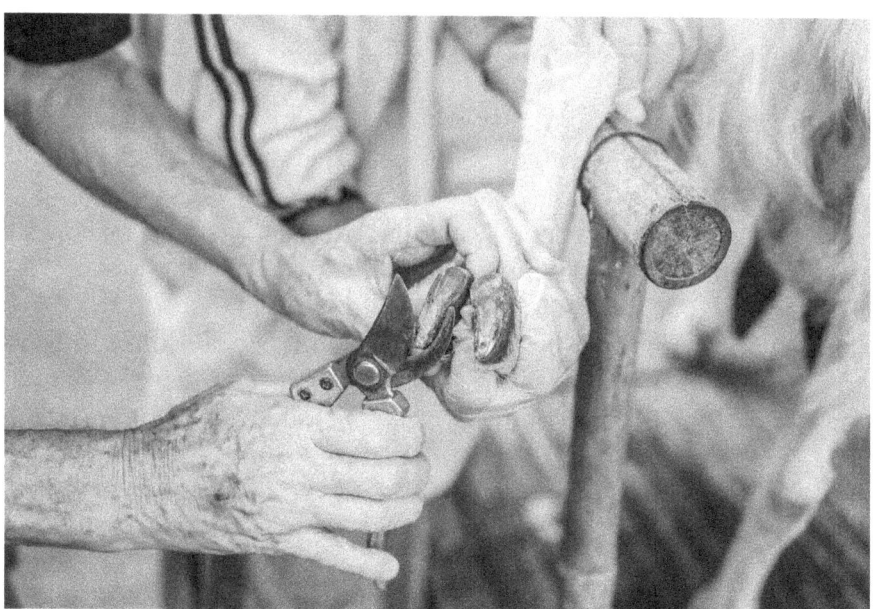

open access to a rough and rocky landscape. They have soft, grassy pasture lands that are inadequate for naturally filing down hooves.

In our pasture, there is an old fieldstone foundation from a former outbuilding that was torn down long before we bought the property. Our goats love to climb all over it, and we appreciate the fact that it helps wear down their hooves. It is not enough to completely solve the problem, though. We still have to trim their hooves periodically, especially before county fair time. Judges look at how well-groomed and cared-for the goats are.

Left untrimmed, a goat's hooves can become overgrown, leading to pain, difficulty walking, and poor posture. Overgrown hooves can also lead to medical problems like abscesses, hoof rot, and foot scald, a term for inflammation of the tissue between the goat's toes.

You will need a few tools to help you trim your goats' hooves, including a pair of scissor-like hoof trimmers, a coarse metal file, a hoof pick, and a wire brush. Find a clean, well-lit area in which to work and recruit a helper or two to keep the goat secure. We have found that if you offer the goat a bowl of feed or grain, it will usually be too distracted with the treat to put up a fight.

Start by cleaning off the goat's hooves with a pick or brush. Matted straw or bedding, manure, and grass can get stuck under the cracks and crevices of the hooves. Grasp a hoof and inspect it. You should notice overgrowth along the front and sides of the hoof walls, which you can trim off with the hoof trimmers. Be careful to only cut the hard hoof material, not the soft, fleshy parts that contain the nerve endings and blood vessels. If you accidentally nick this area, apply antiseptic powder to stop the bleeding and clean the wound.

Once the bulk of the overgrown material has been cut away, use the trimmers to remove any jagged or uneven cuts. Then, use the file to smooth and shape the edges of the hoof. Eliminating sharp edges will reduce the chances of discomfort or injury. Move on to the next hoof until the goat's feet are all neatly trimmed.

Dewclaw Maintenance

Many goats have dewclaws, small, non-weight-bearing, extra hoof-like appendages located on their back legs above the hooves. They look like small, curved spurs. These are vestiges of an evolutionary development that, according to some biologists, may have been used to help goats maintain their balance while climbing. Nowadays, the dewclaws serve no real purpose. Most of the time, especially with younger goats, dewclaws require no maintenance or care on your part. However, that's not always the case with older goats. As a goat ages, the dewclaws can become longer, drier, and harder. They can snag on fencing or branches and cause pain and discomfort. If you have some older goats on your hobby farm, pay attention to their dewclaws. You may need to trim those when you trim the goat's hooves.

Using your hoof trimmer, simply trim down the dewclaw. Be careful not to trim it so close to the skin that it bleeds. Then, file it so it is smooth and rounded. It is a quick and easy task, but one that will make the goat feel more comfortable.

Coat Care

Goat fur should be soft and luxurious. This is especially true of goat breeds that are raised for their fibers; however, all goats will have thick, downy coats if you ensure that they are fed a nutritious diet with the minerals they need. Fortunately, maintaining your goats' fur is not time-consuming or labor-intensive. Brush your goats from time to time with a firm-bristle grooming brush. You can find these online or at your local feed store. When you brush your goat, you will be removing loose fur, dandruff, dirt, and debris while stimulating blood flow along the animal's skin.

Goats don't need to be brushed often. A few good brushings in the spring to remove the winter debris and a few more in the summer to remove the excess hair are really all that is needed, but if you are raising your goats for their fibers, you will want to spend more time grooming their fur. Since my children showed their goats at the county fair, we also made sure to brush them more frequently to keep them looking sharp.

We also bathed our goats right before county fair time. Goats don't require bathing and washing, but we knew that the goat show judges wanted to see clean, well-groomed goats. We used a mild dog shampoo and a garden hose to get the job done. Since our fair is held every July, during the hottest part of the summer, the goats didn't mind the cold hose water. In fact, I think they enjoyed it. We never used a blow dryer to dry our goats' fur after bathing them, but I know people who did. If you want to dry your goat's fur, just be careful to keep the blow dryer far enough away from the animal that you don't burn its skin.

Depending on the breed of goat you have and the purpose of your goats, you may want to add "haircuts" to your list of annual tasks for your herd. We clipped our goats once a year. Again, this was done in preparation for the county fair. Whether you plan to show your goats or not, a summertime haircut for your goats is not a bad idea. It will help them stay cooler in the summer heat. We use an electric trimmer/shaver set at number two. If you have a goat breed with sun-sensitive skin, you may want to avoid clipping their fur altogether as that protects their skin from sunburn.

If you are milking your goats, you will want to routinely shave the hair from the goat's underside and udders. It will be easier to clean the area

Photo Courtesy of Caroline Huntley Pondicherry Farms

before you start milking and will prevent hairs and debris from falling into the milk.

Bathing, brushing, and clipping your goats' fur is also a good time to inspect their coats for external parasites. Goats are susceptible to goat lice, tiny, tan-colored insects that invade the animals' fur. If a goat has a dull, splotchy coat, and you see it frequently scratching itself on fence posts and trees, it could be a sign of goat lice. Do a close inspection of the animal's fur to see if you can spot the small bugs or their gray eggs. Fortunately, goat lice is easy to treat using louse powder. Your veterinarian can recommend a brand for you to use. If you find lice on one goat, treat the entire herd with the louse powder. Also, remove and replace all the bedding and thoroughly clean the goat barn to prevent re-infestation.

Record Keeping

Develop and maintain a record-keeping system to help you keep track of your goats' health history. You can use a journal, a notebook, or a spreadsheet on your computer—whatever method you are most comfortable with. Document everything, including what vaccines were given, when and why deworming was done, breeding information, the dates you trimmed hooves, and any additional health issues or medical treatments that arise. Don't trust your memory to recall important dates. A record-keeping method is more accurate and complete.

Maintaining Optimal Diet and Nutrition

Chapter 7 of this book is dedicated to the proper feeding and nutritional requirements of goats; however, since diet plays an integral role in the health of your hobby farm goats, we would be remiss if we didn't mention it here. Instead of discussing the actual feed itself, which you can read about in Chapter 7, let's discuss the peripherals of feeding your goats to help you better understand the steps you can take to maintain optimal nutrition for your herd.

Foraging

As born foragers, your goats will spend a large portion of their day browsing their pasture for tasty foliage to snack on. This represents a percentage of your goats' diet that is out of your control. Or is it? As a hobby farmer, you have the ability and responsibility to make sure that your goats have access to a variety of quality plants. Inspect your pasture area on a regular basis to look for diseased trees or plants. There may be mold, dust, black spots, spider webs, or shriveled leaves. Remove these plants as soon as possible and burn them to prevent the spread of plant diseases or insect infestations.

Remove all trees, bushes, and plants that are toxic to goats, like black cherry trees, lily of the valley, foxglove, milkweed, boxwood, and elderberry, to name a few. While most of these won't kill your goats, at least not in small amounts, they may sicken them. All plants that pose a negative threat to your goats' health should be removed from the pasture. Ideally, you will have the ability to rotate pastures, although that can sometimes be a challenge for hobby farmers with limited space. By moving your goat herd to a different pasture, you have a chance to reseed the vacant one with plants that are nutritious for your goats. Red and white clover, alfalfa, tall fescue, Italian ryegrass, and chicory are all good options. Better yet, buy a seed mixture that contains all of these. The variety of fast-growing plants will give your goats new, fresh, and tasty greens to snack on.

Sanitation

The quality of the food you provide for your goats can be undermined if the feed dishes are not clean. Uneaten feed can spoil, get moldy, or attract insects, all of which will taint the next scoop of feed you pour in. Get in the habit of discarding old feed remaining in the dish and thoroughly washing the feed dish. Do the same with the goat waterer, too. Good sanitation is key to maintaining the health of your herd, so take a few minutes each day to make sure the food and water containers are clean.

Photo Courtesy of Mary Beth O'Donovan

Seasonal Variations

The nutritional needs of your goats may vary with the changing seasons and weather conditions. You will need to adjust your feeding habits accordingly to ensure that your goats are getting the nutrients they need all year long.

Invest in Quality Hay

I can't tell you how many times I've had hay farmers offer to sell me "discount" hay, which they call "goat hay," when I've inquired about purchasing bales of hay for my goats. When a hay farmer uses the term "goat hay", what he or she is really saying is that the hay has gotten moldy

HEALTH ALERT
Listeriosis

Listeriosis, also known as circling disease and silage sickness, is an infection usually caused by feeding goats moldy hay. This infection affects the goat's brainstem and is life-threatening. Early signs of listeriosis include loss of appetite, decreased milk, and fever. Advanced symptoms include disorientation, drooling, seizures, and partial paralysis. Early administration of medication is recommended to prevent death from listeriosis. Good hygiene and responsible feed storage can help reduce the risk of this disease.

and is, therefore, unsuitable for horses. Hay needs to be as dry as possible prior to baling to prevent the growth of mold and mildew. Horses are sensitive to moldy hay and either refuse to eat it or will become sick from eating it. A hay farmer doesn't want to get stuck with a bunch of moldy hay that horse owners won't buy, so they offer it at a discount price to unsuspecting goat farmers. To be clear, it is not that hay farmers want to get your goats sick. Most of them believe the myth that goats will eat anything, and they are trying to unload bales of hay that have lost their value. When buying hay for your goats, be sure to ask a lot of questions about the quality of the hay and if it has been exposed to moisture. Don't settle for goat hay.

Secure Food Storage

Hay and goat feed should both be stored in moisture-proof and rodent-proof containers and in an area that is off-limits to your goats. Our first goat barn did not have a secure feed room. There was just a four-foot-high wall separating the goat area from the storage area. We learned pretty quickly that goats are determined problem solvers, especially when they get the munchies. A few of the goats figured out how to jump the wall, pull off the lid to the feed bin, and help themselves to a feast. We had a couple of miserably bloated goats—kind of like me after Thanksgiving dinner. After a few of these chow sessions, we moved the feed into the garage. The new goat barn has a dedicated feed room with a door that securely locks. Even though the key is hanging right in plain sight, they haven't figured out how to use it to open the door. Yet.

Identifying Health Concerns

In Chapter 5, you will be presented with information about the most common goat diseases, as well as the symptoms and treatment plans for each of them. As part of the overall health and upkeep of your hobby farm goat herd, you should be able to identify signs and symptoms that could point to a health concern or injury. Some of these symptoms include the following:

- Diarrhea
- Constipation
- Bloody stool
- Bloody urine
- Abnormal breathing
- Coughing
- Wheezing
- Vomiting
- Nasal discharge
- Eye discharge
- Restlessness
- Loss of appetite
- Drooling
- Self-isolation
- Lethargy
- Excessive vocalization
- Refusing to drink
- Excessive thirst
- Weight loss
- Excessive weight gain
- Bloated or distended abdomen
- Swelling or bleeding anywhere on the body
- Paralysis
- Refusing to stand
- Limping
- Hair loss

These signs and symptoms are general and can indicate a wide range of health concerns or illnesses. If you observe any worrisome symptoms such as these, consult with your veterinarian. He or she can provide you with an accurate diagnosis and set up an appropriate treatment plan. While routine veterinary check-ups, good hygiene practices, and preventative care are crucial for maintaining the health and well-being of your hobby farm goat herd, things can still go wrong. Your goats can become sick or injured. If that happens, the earlier you can recognize a problem, the sooner you can start to treat it.

CHAPTER 5

Goat Diseases and Preventive Measures

One summer, we hosted a big family cookout at our hobby farm. There were more than fifty people, including a bunch of my young nieces and nephews—none of whom live on a hobby farm. The youngsters were fascinated by the goats, as well as our chickens, ducks, and rabbits. My husband was busy grilling burgers and corn on the cob, and I was busy setting up the food table with potato salad, sliced watermelon, and all the other typical side dishes that are common at summer parties. We were both trying to make sure the party went smoothly and trying to talk to all our guests. Admittedly, we were both too busy and too distracted to see what was happening in the goat barn.

The next morning, two of our goats, does Ruby and Callie, remained in the goat barn when the others went out to the pasture. They were lying on their sides, bleating in discomfort and salivating. When we got them standing up, we could see that they both had distended abdomens and were walking funny. We contacted our veterinarian, who promptly came out to see them. She concluded that Ruby and Callie were suffering from bloat, a buildup of gas in the rumen, the first of the four stomach compartments.

The veterinarian questioned us about the food the goats had recently eaten. Only then did our youngest daughter confess that she and her young cousins had fed Callie and Ruby *all* the watermelon rinds and the leaves from the ears of corn from the family cookout. They also gave them a few too many scoops of their goat feed, leaves from the nearby maple tree, and Lord knows what else from the cookout. Apparently, the kids were playing "farmer." The sudden influx of all these moisture-rich foods in the goats' digestive system caused a buildup of painful gas.

Bloat can be deadly, but fortunately, our vet gave the two goats some medication to alleviate the gas and help them digest the gluttony of treats they were fed. Ruby and Callie recovered in a few days, and our daughter got a big lecture. We explained to her why the goats got sick and told her that, in the future, she needed to have an adult accompany her and her young cousins or friends to the goat barn. My husband and I also learned that we need to restrict access to our animals when we have guests over and that we have to more closely monitor what is going on in the goat barn and pasture during these times.

The bloat incident aside, we have been lucky with our hobby farm goat herd. They have enjoyed relatively good health, though we have lost a few goats here and there from illness. Goats, like other livestock animals, can be susceptible to various diseases that can negatively affect your herd's health. Goats are generally hardy animals, but there are some illnesses that are particularly linked to goats. The prevalence of illness depends on other factors, along with bacteria, viruses, and pathogens, including living conditions, genetics, and contact with other animals.

Photo Courtesy of Mary Beth O'Donovan

In this chapter, we will cover several of the most common goat illnesses and diseases. You will learn the signs and symptoms of each one, as well as the causes and treatment options. Chances are quite high that you will only encounter one or two of these diseases on your small-scale hobby farm; however, you should have an understanding of the potential disease threats and how to prevent them.

Coccidiosis

A common intestinal parasitic disease, coccidiosis is caused by the coccidia protozoa. When these parasites invade the goat's intestines, the result is inflammation of the lining of the intestines, damage to the intestinal walls, and issues absorbing key nutrients. Although goats of all ages can contract coccidiosis, it is more prevalent in younger goats. The type and severity of the symptoms of coccidiosis can vary depending on the age and relative health of the goat and the type of coccidia parasite.

The number one symptom of coccidiosis is diarrhea. The diarrhea will be watery and may appear to contain mucus or blood. As a result of the diarrhea, the goat can quickly become dehydrated. A dehydrated animal will be lethargic with sunken eyes and a dry mouth and nose. Goats infected with coccidiosis will also have poor appetite and may refuse to eat, leading to weight loss. The weight loss can also be a result of the lack of nutrient absorption in the animal's intestines.

The parasite that causes coccidiosis spreads from one goat to another through feces. A goat infected with coccidiosis will shed the parasite's eggs

Coccidiosis

in its poop. If that poop contaminates the food, water, or pasture and another goat ingests it, it will also become infected by the parasite. Proper sanitation and hygiene practices can help prevent the spread of the parasite. Thoroughly washing the food and water containers on a regular basis, keeping the goat barn pens clean, and properly disposing of manure will greatly reduce the spread of coccidiosis. That includes raking and removing the nannyberries from the pasture as well, but these efforts won't completely eliminate the problem. You simply can't be in the pasture all day, every day, removing every nannyberry as soon as it hits the ground.

Early diagnosis and treatment of coccidiosis are key to the goat's recovery. Your veterinarian will probably prescribe one of several anti-coccidial medications that are available, such as amprolium, tol-trazuril, or sulfadimethoxine. The specific medication the veterinarian prescribes, as well as the dosage and duration of treatment, will depend on the age of the sick goat and how severe the infection is.

Your veterinarian will also advise you to isolate infected goats to prevent them from spreading the disease to the healthy goats in your herd and to reduce the likelihood of the goat barn becoming contaminated. The sick goats should have access to fresh, clean water that is replenished throughout the day and good-quality, easy-to-digest foods. You will need to work to ensure that the sick goats maintain an adequate hydration level so their conditions do not worsen.

If you experience frequent coccidiosis outbreaks on your hobby farm or you live in an area where this disease is more common, speak to your veterinarian about preventative medications. There are some medicines on the market that interrupt the life cycle of the coccidia protozoa. There are also goat feeds that contain anti-coccidia additives. Because there is the risk of the protozoa becoming resistant to medications, it is always best to discuss preventative medications with your veterinarian and follow his or her recommendations regarding using these products.

Also, discuss any all-natural, homemade treatments for coccidiosis that you want to try with your veterinarian to make sure that the substances you use are safe for your goats, won't counteract the medications they are on, and to gauge the effectiveness of the home remedies. For example, you may hear experienced, sustainable goat farmers swear by oregano essential oil to prevent coccidiosis. While oregano does have antibiotic properties, it can also damage a goat's digestive system, especially if it is used indiscriminately. You may also be told that apple cider vinegar is an effective treatment option. The acidity in the vinegar can kill the protozoa, but it is extremely harsh and can burn the goat's intestines. A better use of apple cider vinegar is as a cleaning agent. Spraying a mixture of apple cider vinegar and water in the corners, nooks, and crannies in the goat pens might kill the protozoa eggs that you missed while cleaning.

Bloat

As I mentioned at the beginning of this chapter, the accumulation of excess gas in the goat's rumen, the first of its four-part stomach, causes a condition known as bloat, or more technically, ruminal tympany. What I didn't mention earlier is that there are two different kinds of bloat. The one that our goats had was frothy bloat. This occurs when the animal's rumen fills with foamy, frothy goo that prevents the release of gas. As we learned, frothy bloat is caused by the consumption of moist, fermentable food, like fresh clover or, in my case, corn cob leaves and watermelon rinds. The goat's rumen produces gas as it breaks down the fermentable

Bloat

plant material. Since the gas cannot escape, the rumen distends, causing discomfort and potentially life-threatening complications.

The other type of bloat, free gas bloat, is similar to frothy bloat in that gas is trapped in the goat's rumen, but in this condition, the blockage is caused by a foreign object or some other form of obstruction in the goat's esophagus.

Goats suffering from bloat will have a noticeably swollen and tight abdomen on the left side. The animal will show signs of being in pain and discomfort. As the rumen expands, it can put pressure on the goat's diaphragm, leading to difficult and labored breathing. The animal may be restless, frequently standing up and lying down, or it may be lethargic and refuse to stand. It will also refuse to eat and may even refuse to drink water.

Bloat has the potential to be quite serious, even life-threatening, so you should contact your veterinarian as soon as possible. The sooner you can start treatment for bloat, the better it will be on your goat. Your veterinarian will probably do what our veterinarian did—administer an anti-bloat medication that contains ingredients that can break up the foam and help the gas escape from the rumen. Poloxalene and simethicone are two of the most commonly used ingredients found in anti-bloat medication.

If the goat's condition is dire, the veterinarian may insert a stomach tube into the goat's rumen to quickly release the trapped gas. Don't attempt to do this procedure on your own. You could cause more harm. Your veterinarian has the knowledge and experience to safely and effectively do the procedure.

There is another procedure your veterinarian might do, but it is usually reserved for life-or-death cases. A trocarization is a surgical procedure in which an incision is made through the skin near the rumen. A tube is inserted through the opening and into the rumen to release the gas.

Recovery from bloat involves keeping the goat hydrated and ensuring its electrolytes are in balance. If a trocarization was performed, you will need to monitor the incision site and do basic wound care. Watch the goat's diet for the next several days and observe the animal's behavior. Contact your veterinarian if you notice any concerning new symptoms or worsening of existing symptoms.

Caprine Arthritis Encephalitis

Caprine arthritis encephalitis, or CAE, is a viral disease that impacts the central nervous system and joints of an infected goat. The virus can be passed from one goat to another in three different ways. First, the virus can pass from an infected mother to her kid through her milk. The virus can also spread from goat to goat via close contact and shared body fluids, including during mating and through a goat's sneeze or cough. Lastly, the CAE virus can be spread through contaminated equipment, including surgical tools, vaccination needles, or tattooing equipment. The virus can remain on equipment that hasn't been thoroughly sterilized.

The second two words of the CAE name offer a clue about the symptoms of this disease. The "A" stands for "arthritis," and CAE can cause chronic inflammation of the joints, stiffness, swelling, and limited mobility. The goat's affected joints might also be warm, and the animal will react in pain to your touch.

The "E" in CAE stands for "encephalitis," a term that means swelling of the brain. The CAE virus can manifest in issues with the animal's central nervous system and brain. Symptoms can include seizures, loss of coordination, muscle tremors, erratic behavior, and paralysis.

In addition to these symptoms, a goat infected with CAE may experience weight loss, respiratory issues, a decrease in milk output, and a general health decline.

Unfortunately, there is not yet a cure for CAE. An infected goat carries the disease throughout its life, and the virus becomes part of its genetic makeup. The best treatment option is to take steps to control the spread of the disease. Many small-scale goat farmers opt to identify the goats in their herd that test positive for carrying the CAE virus. Those goats are culled to prevent the rest of the goats from becoming infected.

As soon as you suspect a goat could have CAE, remove it from the herd and isolate it from other animals. This is an important step to prevent direct transmission. You can also reduce the chance of kids becoming infected with the virus from their mother's milk by pasteurizing the milk using heat to kill the virus. Once it has cooled, you can use the pasteurized milk to bottle-feed the kids. By maintaining strict biosecurity measures and hygiene practices, you can also help prevent the spread of the CAE virus. That includes properly disinfecting all equipment, feed containers, and housing.

Urinary Calculi

A type of urolithiasis, urinary calculi are stones in the goat's urinary tract. They are similar to kidney stones in humans. The animal's body secretes a mineral substance that becomes solid, crystal-like stones that can block the urinary tract. Although both male and female goats can develop these stones, they are more common in male goats, both bucks and wethers, simply because their internal anatomy is more conducive to blockage.

The urinary stones are primarily caused by insufficient water intake and imbalances in the goat's diet. The stones are composed of calcium and phosphorus and occur when the balance of minerals is off. The chances of developing urinary calculi increase if the goat is not drinking enough water. The water flushes the calcium out of the goat's system so stones can't form. Another factor that contributes to the development of urinary calculi is diet. Goats that are fed a diet that consists mostly of grain feed and a comparatively lower amount of dry roughage, like hay, are more apt to suffer from urinary stones. Like water, roughage helps push everything through the goat's system so that minerals cannot linger long enough to crystallize into stones.

If you notice a goat in your hobby farm herd that seems to be straining to urinate or assumes the urinating position and nothing comes out, it could be that urinary stones are causing a blockage. Other signs and symptoms of urinary calculi include blood in the urine, pain or discomfort, and a distended lower abdomen.

Left untreated, urinary calculi can be fatal in goats. The stones can completely block the urinary tract, leading to kidney and bladder failure. Urinary calculi are a medical emergency, and immediate veterinary intervention is necessary to remove the obstruction. Depending on the severity of the goat's symptoms, your veterinarian may try to open the urinary tract by inserting a catheter into the goat's urethra to direct the urine around the stone and give the goat some relief from its discomfort.

Another treatment option is to pump the goat full of fluids via an IV to correct the animal's hydration levels. In milder cases of urinary calculi, this can be enough to dislodge the stone and flush it out. If the goat is in grave condition, it may need surgical intervention to remove the stone, restore the flow of urine, and repair any damage to the urinary tract.

During the goat's recovery from urinary calculi, the veterinarian will ask you to closely monitor the animal's diet to make sure it is getting the correct balance of minerals (particularly calcium), phosphorus, and magnesium. The goat's diet should also have a high level of roughage content to promote proper digestion and waste elimination. You will also be asked to provide the recovering goat with plenty of clean, fresh water. Hydration is key to preventing urine concentration. If surgery was required, you will need to make sure the incision site stays clean to avoid infection. Seek prompt medical assistance if the goat's symptoms return or its condition worsens.

Caseous Lymphadenitis

A chronic bacterial infection that impacts ruminants like goats, caseous lymphadenitis, known as CL, is caused by the bacterium Corynebacterium pseudotuberculosis that enters the goat's body through an open cut or wound or by absorption into a mucus membrane. The infection targets the lymph nodes, resulting in the formation of abscesses.

A goat contracts CL through direct contact with the bacteria. Infected animals can shed the bacteria through milk, nasal secretions, or draining abscesses. If another goat comes into contact with these bodily fluids, the bacteria can enter its body. When an infected goat eats from a shared feed container and drinks from a common waterer, it can contaminate the food and water that other goats in the herd consume. Eating contaminated food or drinking contaminated water can introduce the bacterium into the goat's body.

Goats can also become infected with CL from contaminated farm equipment or living quarters. The bacteria can survive on surfaces for months, even up to a year, and infect your herd long after you think the problem is over. All tools, equipment, housing areas, feed and water

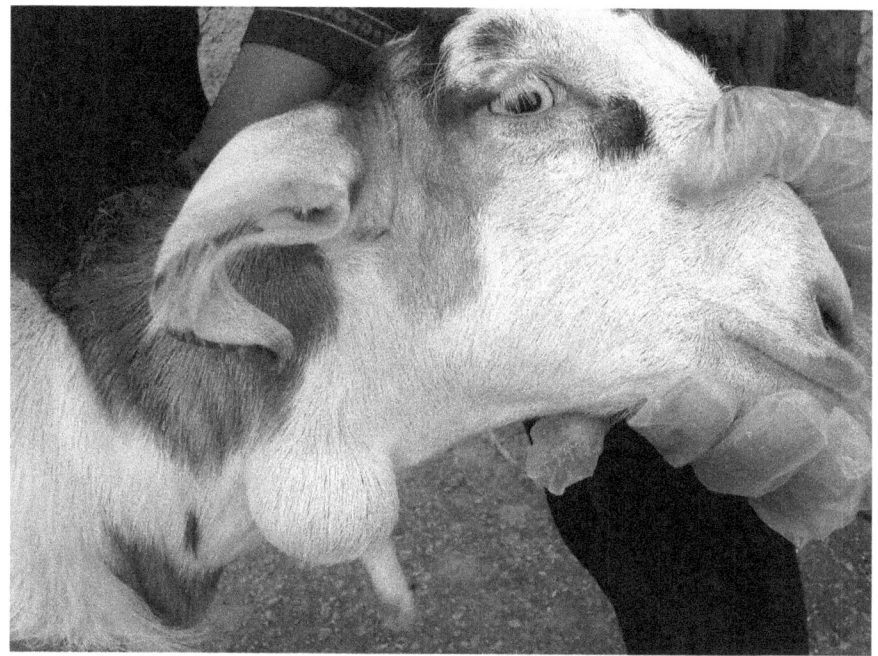

Caseous Lymphadenitis

dishes, and more should be thoroughly cleaned and sanitized to completely kill the bacteria.

The most common sign of CL is the growth of firm, localized abscesses under the goat's skin. They often develop in the lymph nodes, so you will notice them around the animal's neck, head, chest, and rear. The abscesses can, however, form on other parts of the goat's body. The affected lymph nodes may become enlarged, swollen, and tender. Some of the abscesses can remain small while others grow larger. They might even rupture and ooze thick pus.

A goat suffering from CL may have a reduced appetite and lose weight. Does may exhibit a decline in milk production. Since CL is a chronic condition, the overall health of the goat will decline over time. In rare cases, the animal can develop abscesses on its liver, lungs, or other internal organs.

Caseous lymphadenitis is a stubborn bacterium that is difficult to eliminate completely. The treatment of CL is primarily focused on managing the symptoms and reducing the spread of the infection to other members of your goat herd. In some cases, the goat may need to be culled to prevent spreading the infection to other animals.

If you and your veterinarian decide to manage the goat's symptoms, the veterinarian will prescribe antibiotics to control the spread of the infection and treat secondary infections. You should note, however, that antibiotics cannot completely eliminate the CL bacteria from the goat's body. Your veterinarian may also treat the CL symptoms by surgically draining and thoroughly cleaning the abscesses.

The spread of caseous lymphadenitis involves adopting strict hygiene and preventative practices. Quarantine new goats before you integrate them into your hobby farm herd until you can determine their CL status. Maintain a clean and hygienic farm. Regularly disinfect the housing area, equipment, and shared food and water dishes to reduce the risk of transmission. Work with your veterinarian to routinely test your goats for CL so you can quickly identify infected animals and remove them from your herd.

Enterotoxemia

Although enterotoxemia is also called "overeating disease," it is a bacterial infection. The Clostridium perfringens type D bacterium produces toxins that impact the intestines and other internal organs of the goat, leading to enterotoxemia. It is contracted by ingesting the bacteria or its spores. It is called overeating disease because it can come on after the goat overindulges on highly fermentable foods, like grain or fresh pasture greens. The bacterium, which is found naturally in the soil, finds an ideal growing environment in the goat's digestive tract after it has consumed high amounts of fermentable foods.

Goats that have a weakened immune system or are experiencing increased levels of stress are also susceptible to enterotoxemia because their bodies are not strong enough to fight off the infection. Newly weaned kids, animals that have recently been transported long distances, moved to a new farm, or experienced a traumatic weather event are all susceptible. Honestly, I am surprised my goats didn't contract enterotoxemia after the tree fell on their goat barn. That would have seriously rattled me!

In acute cases of enterotoxemia, the goat can die suddenly without exhibiting any symptoms. In most other cases, however, you will observe symptoms. They can include lethargy, weakness, lack of appetite, and low energy. The goat could be experiencing abdominal pain and discomfort that manifests in stretching, standing hunched over, and attempting to kick its stomach with its hooves. Diarrhea is also common in enterotoxemia. The feces are typically watery and loose. Lastly, an infected goat could have neurological symptoms, such as seizures, convulsions, tremors, and spasms.

Treatment for enterotoxemia in goats often involves supportive care and appropriate medications. The treatment that your veterinarian chooses will depend on the severity of the disease. It may include broad-spectrum antibiotics to mitigate the infection and prevent further complications. The veterinarian may also prescribe pain medication to make the animal more comfortable. Fluids, including electrolytes and glucose solutions, can also help the goat stay hydrated and maintain a healthy balance of electrolytes and minerals. In severe cases of enterotoxemia, your veterinarian may recommend using an antitoxin. This substance is derived from plasma containing antibodies that can neutralize the toxin.

Preventative measures are recommended to reduce the risk of enterotoxemia in your goat herd. There are, for example, commercially available vaccines that help stimulate the animals' natural immunity against the toxins produced by the bacteria. As a small-scale goat farmer, you can take steps to reduce stress factors and control the diets of the animals to avoid their overeating of green, highly fermentable foods.

Scrapie

Scrapie is the goat and sheep equivalent of chronic wasting disease that impacts deer populations. You hear about chronic wasting disease in the news from time to time. The disease is a type of transmissible spongiform encephalopathy caused by the accumulation of abnormal prion proteins in the animal's central nervous system. The most common form of transmission happens when an infected doe passes the disease to her offspring during birth, as the amniotic fluid contains the prion proteins, but goats can also come in contact with the prion particles through contaminated feed, equipment, tools, or housing.

As a neurological disease, scrapie is progressive and degenerative. There is no cure for scrapie, and the disease will lead to the goat's death. Scrapie management strategies focus mainly on preventative measures to ensure hobby farmers can maintain a scrapie-free herd. Every summer, prior to county fair time, we had to make arrangements for our goats to be tested for scrapie and certified as scrapie-free. This was required of all goats and sheep before they were allowed onto the fairgrounds. I believe this is standard practice in many states because it is a good way to identify sick animals and prevent them from coming into contact with other animals. To my knowledge, none of the goats at our county fair ever tested positive for the disease, but I support strict biosecurity controls.

Stopping the spread of scrapie requires goat farms to limit contact with other herds. Newly acquired goats should be quarantined from the rest of the herd until they can be tested for scrapie. When purchasing

Scrapie

new goats for your hobby farm, ask if the farm is certified as scrapie-free before you make your buying decision.

The symptoms of scrapie can vary depending on how far the disease has progressed. The goat may suffer from intense itching. You might notice the animal scratching or rubbing against things more than normal. The behavior of a goat with scrapie might noticeably change. The animal may be extremely nervous, restless, or act aggressively. Progressive weight loss is also typical in scrapie-infected goats.

As the disease progresses, the goat's coordination will be impacted. You might observe the animal stumbling, falling, walking with an odd gait, or having difficulty walking. Some goats with scrapie exhibit involuntary muscle spasms, tremors, or twitches. Diseased goats might adopt a strange, abnormal posture. They might arch their back or walk with their heads down.

Many states, regions, and countries have scrapie control programs in place in an effort to manage the disease and prevent its spread. Your veterinarian will want to regularly test your goats for scrapie. Unfortunately, since there is no cure for the disease, infected goats and their offspring should be culled to stop the spread of scrapie.

White Muscle Disease

Also called nutritional myodegeneration and selenium deficiency, white muscle disease is a condition that affects goats, as well as other livestock animals. It is caused by a vitamin deficiency, specifically a deficiency in either selenium or vitamin E or both; these are essential nutrients for muscle health, growth, and function.

Selenium is a trace mineral, while vitamin E is a fat-soluble vitamin. Both of these play vital roles in the body's defense system and in muscle health. If the goat's diet is lacking in selenium, vitamin E, or both, it can be detrimental to the animal's health. Goats that are fed primarily by grazing and have a pasture that lacks plants rich in vitamin E or selenium may not get sufficient amounts of the nutrients. This can also happen if their grain diet contains an imbalance of minerals. Lastly, the deficiency can

White Muscle Disease

be the result of poor absorption of nutrients in the goat's digestive tract caused by an intestinal disorder.

Goats suffering from white muscle disease exhibit generalized muscle weakness and stiffness. The animals may be reluctant to stand up, have difficulty walking, have limited range of motion, and walk with a stiff, unnatural gait.

Treatment of white muscle disease in goats involves correcting the deficiency in vitamin E and/or selenium. Under the supervision of your veterinarian, the goat will be given supplements, either orally or through injections, to bring its levels of selenium and vitamin E back up to normal. The exact dosage, method of administering, and duration of supplement treatment will all depend on the severity of the deficiency.

Preventing white muscle disease is fairly simple. It requires providing a balanced diet that meets the nutritional requirements of goats and includes the recommended amounts of selenium and vitamin E. It may involve enhancing your pasture to include high-quality forage, offering mineral blocks or supplements, or ensuring the commercial goat feed is fortified with selenium and vitamin E.

Goat Pox

You have heard of chicken pox, which affects humans—not chickens. But did you know there is goat pox, too? Don't worry. Unlike chicken pox, goat pox does not affect humans. Goat pox is a viral disease caused by the Capripoxvirus, which results in skin lesions and fever.

Goat pox is primarily contracted through direct contact with infected animals or through exposure to tools, bedding material, and feed containers that have been contaminated with the virus.

Classic skin lesions are the hallmark of goat pox. They develop as raised, scab-like sores on the various parts of the goat's body, including the face, mouth, genital area, and udder. The lesions start out as small red bumps but quickly progress to larger, fluid-filled pox that eventually scab over. In addition to the skin lesions, the goat may have a fever, loss of appetite, and reduced milk output.

Got Pox

There is no specific antiviral treatment available for goat pox; the treatment options focus on supportive care and managing the symptoms while the goat recovers. That includes maintaining proper hydration, ensuring good nutrition, and minimizing stress. You can also apply topical ointments to the skin lesions to reduce the risk of secondary bacterial infections. The prognosis is good for goats suffering from goat pox.

While there is a vaccine you can use to prevent an outbreak of goat pox, consult with your veterinarian to determine if it is in the best interest of your hobby farm goat herd to use the vaccine. You can reduce the chances of your goats contracting goat pox by limiting contact with outside animals, quarantining new goats before you integrate them into your herd, and thoroughly cleaning all facilities and equipment.

Foot Rot

Foot rot in goats is just as unpleasant as the name suggests. It is a bacterial infection of the animal's hooves, particularly the fleshy area between the toes. It is caused by several different bacteria, but the main culprits are often Fusobacterium necrophorum and Dichelobacter nodosus.

Bacteria thrive in wet, dirty, muddy conditions. And prolonged exposure to wetness can soften goat hooves, making them more susceptible to bacterial infections. In addition to contracting foot rot from poor environmental conditions, it can also be spread through direct contact with goats that carry the bacteria.

Foot rot can cause a goat to favor the infected foot. It will limp or hold one foot off the ground. You may notice that the area between the goat's toes is red, inflamed, and swollen. It might even have a foul odor and a pussy discharge. If the infection becomes severe, it could cause the hoof wall to separate from the underlying tissue, leaving painful, open sores.

A few simple, preventative measures can greatly lower your animals' risk of getting foot rot. The first is to keep the pens or goat barn as clean and dry as possible. Stagnant water mixed with manure, urine, mud, dirt, and soiled bedding is the perfect breeding ground for bacteria. When goats stand in this contaminated puddle for a long time, their hooves will soften. This gives bacteria a literal foothold to enter the goats' feet and

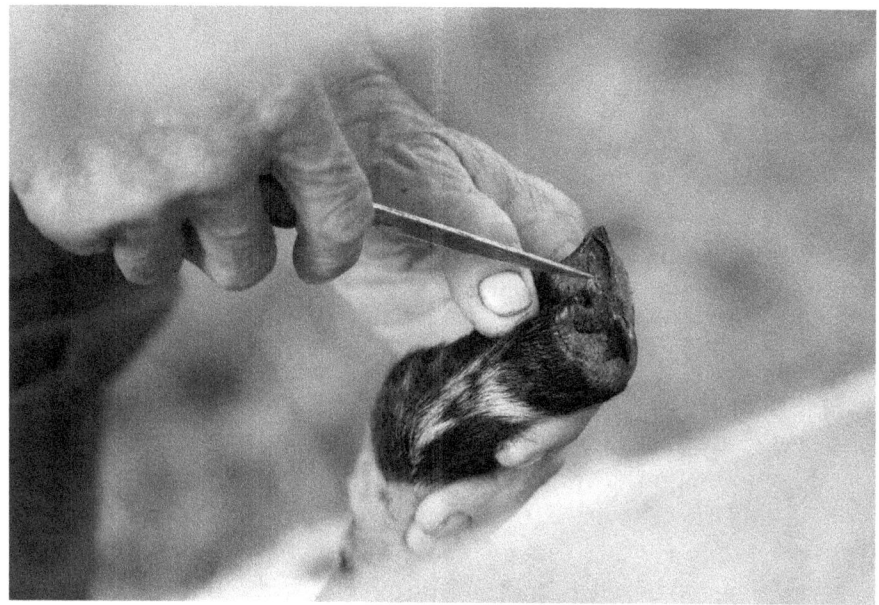

Foot Root

cause infections. In addition to adopting good hygiene practices, be sure you are taking good care of your goats' hooves. Regular hoof trimming will keep their feet in good condition and stave off infection.

Foot rot requires medical attention. One option is to soak the infected hooves in a foot bath solution to kill bacteria. Discuss this with your veterinarian. He or she may recommend a solution that includes zinc sulfate, copper sulfate, or formalin. Applying topical antibacterial ointments or sprays can help combat infection and speed up healing. Likewise, antiseptic solutions can be used. In stubborn cases of foot rot, your veterinarian may prescribe a systemic oral antibiotic to kick out the infection once and for all.

During an outbreak of foot rot, you should also trim the infected hooves. When you remove excess hoof material and diseased tissue, you will be better able to access the area to apply ointments. It also allowed the infected area to breathe and heal.

Contagious Ecthyma

A viral infection impacting goats and other ruminants, contagious ecthyma, also known as Orf, is caused by a virus of the Parapoxviridae family. It is spread through direct contact with an infected animal or by exposure to contaminated items like bedding material and food dishes. The disease manifests in skin and mouth lesions that seep fluid. When one goat comes into contact with this fluid, which can happen when goats bed down together, the virus is transmitted.

A classic symptom of Orf is lesions on the goat's lips and mouth. The raised, scab-like sores are painful and can be found on the gums, tongue, lips, and inside the goat's mouth, making eating and drinking difficult and painful. In addition to the oral lesions, contagious ecthyma can cause skin lesions on other parts of the body, including the ear and udders. The lesions itch so much that the goat will scratch and rub against rough objects in search of relief. Doing so, however, rips open the lesions, putting the animal at risk for secondary infections.

In most cases, contagious ecthyma is a self-limiting disease. Infected goats will recover naturally on their own. Treatment really just involves making sure the goat eats and drinks enough and treating the lesions

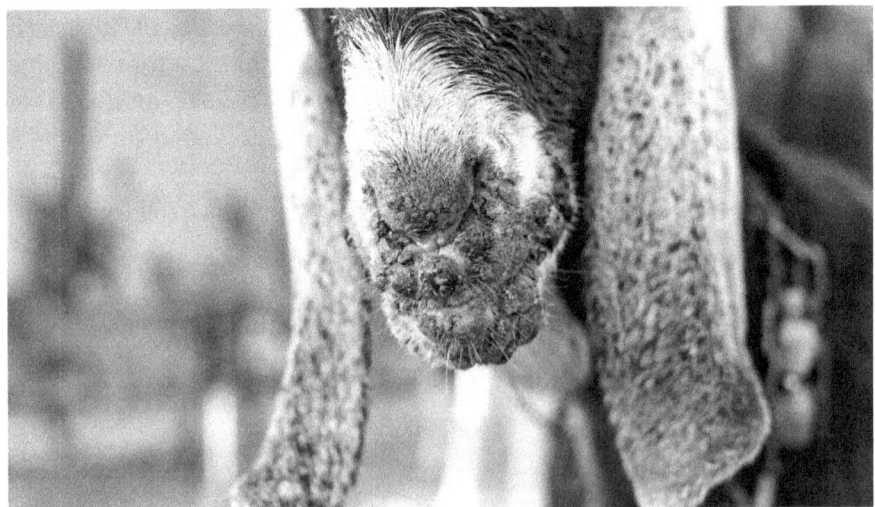

Contagious Ecthyma

with topical antibiotics to prevent secondary infections. I would also recommend that you isolate the affected goat from the rest of the herd, then clean and sanitize the goat barn, feed and water dishes, pens, and other areas to lessen the chance of Orf infecting the rest of your goats.

Implementing Biosecurity Practices to Prevent Disease Outbreaks

You may have noticed that many of the most common goat diseases can be prevented or contained by maintaining good hygiene practices on your hobby farm. Implementing a biosecurity routine is an important step in keeping your goats safe and healthy. Regular and thorough cleaning includes the goats' living area—pens, stalls, or goat barn—as well as all the surfaces and high-touch areas. Milking equipment, water buckets, and feeding troughs should be routinely cleaned and sterilized.

Consider limiting guests and visitors to your hobby farm. The more people that come in contact with your goats, the more chances for

viruses to spread. If you plan to welcome visitors to your farm, ask them to either wash their hands or use hand sanitizer before handling your animals. You can even ask them to wear disposable shoe coverings to prevent the transfer of pathogens from the dirt and manure stuck on the bottom of their shoes.

When you and your family are doing goat chores, use separate shoes or boots that are dedicated to farm chores. I have a pair of slip-on rain boots—Wellingtons—that I only wear when visiting the goat barn.

The whole family got in the habit of wearing farm boots after an incident involving my oldest daughter, a pair of flimsy flip-flops, and a pitchfork. She was a young teen at the time and was one of those girls who only wore flip-flops, no matter the temperature. She was in a rush to do goat chores and ran out to the barn in her flip-flops. One of the goats had knocked over the pitchfork, and, in her hurry, my daughter didn't notice until she stepped on it. One of the tines pierced the bottom of her flip-flop and stuck into the bottom of her foot. She had to sit down and pull it out. Ouch.

I took her to the emergency room because I thought she would need a tetanus shot. I'll be honest. I have no idea where that pitchfork came from or how old it was, but I do know it has a nice patina of rust on it. Plus, you know what that pitchfork was used for and what it regularly touches, right? When I mentioned that to the ER doctor, he told me there is more bacteria in regular soil than in manure. My daughter exclaimed, "That can't be right! I would lick dirt, but I'd never lick poop!" Yikes! Our visit to the ER got her one stitch, one tetanus shot, and a lecture on not embarrassing her mom in front of the doctor. After that, we started a "no flip-flop" policy in the barn.

How to Dispose of Soiled Bedding and Manure

Wet, dirty, soiled bedding could harbor bacteria; therefore, it should be removed and taken far away from the goat barn. Most hobby farmers, especially ones who want to practice sustainability, compost soiled bedding. That's what we do. We did one of those DIY tricks with free wooden pallets and made a compost pile behind the detached garage, partially

hidden by a plump pine tree. The manure and urine are high in nitrates, and the damp straw breaks down fairly quickly.

A neighbor of ours used dirty bedding as fill for a berm he built along one side of his property. He piled up the discarded bedding and manure until it reached the desired height, then he covered it with topsoil and planted grass seed. As the fill material broke down, the size of the berm decreased, but the grass is thick and lush.

You could offer your discarded bedding to goatless friends and neighbors, especially ones who are avid gardeners. They may appreciate having access to compostable manure that they can turn into rich soil for their gardens.

You can contact officials in your community or county to see if they have a place where you can dispose of goat manure. Some rural agricultural areas have such places, but others do not. Most places frown upon disposing of manure and soiled bedding in county dumps or landfills. You just need to find out the rules and restrictions in your area.

Working with a Veterinarian

When you establish your hobby goat farm and welcome your small-scale goat herd, you should develop a relationship with a trusty and knowledgeable veterinarian. View your farm's veterinarian as a partner who will help you keep your goats healthy and thriving. A veterinarian can provide guidance and suggest ways to utilize preventative health measures. Under the veterinarian's advice, you can set up a schedule for vaccinations, routine health checks, and maintenance tasks.

HELPFUL TIP
Sound the Alarm

Goats have a unique way of alerting for danger: sneezing! Although sneezing can indicate a health concern, goats sometimes sneeze to warn their fellow goats of an incoming threat. One theory is that goats sneeze to signal danger as a secret code that only their fellow goats, not a predator, will understand. Other causes for sneezing can include allergies, other irritants, or illness.

When you develop a rapport with a veterinarian, you will have someone to reach out to when you have a sick goat or an unexpected injury occurs. You should get in the habit of observing your goats and noting any signs of illness, such as changes in physical appearance, behavior, or appetite. Should you notice any concerning symptoms, you will be able to seek prompt veterinary assistance. The quicker you can seek medical care, the better it is for your goats.

Quarantine Protocol

One of the most effective preventative health precautions you can implement on your small-scale goat farm is to quarantine all new animals coming onto your hobby farm. When you add a new goat to your herd, keep it in a separate area, away from your established herd. The animal should be isolated for at least 30 days. During that time, you should keep a close eye on the goat. Any signs of illness should be reported to your

veterinarian, who can determine if the animal has a contagious disease that could potentially impact your entire herd.

Quarantining new animals is an important biosecurity measure that should not be skipped or shortened. When you minimize contact between your herd and other goats, livestock animals, and even wildlife, you are reducing the chance that your goats will be contaminated by a disease-carrying animal.

Summary

As a hobby goat farmer, you have some control over the health of your goats. By maintaining good hygiene practices, feeding your goats a nutritious diet, and educating yourself about common diseases, you will be taking important steps toward keeping your goats disease-free. Work with an experienced and knowledgeable veterinarian to develop a comprehensive health management plan—one that includes vaccinations, routine maintenance, and a quarantining protocol—that you can implement on your hobby farm.

CHAPTER 6

Breeding and Reproduction

One spring day, my whole family was visiting Grandma and Grandpa, and on the drive home, we all wondered if Ruby, one of our pregnant does, had delivered her kid yet. According to the date we had circled on the calendar, she was due at any time. As soon as the car came to a stop in the driveway, one of my daughters jumped out and ran to the goat barn to check on Ruby, who was confined in a separate stall. My daughter casually walked back to the driveway and announced in a matter-of-fact tone, "Yep, she had her babies." The rest of us assumed she was joking. But sure enough, a quick peek into her stall confirmed that during the few hours we were away, Ruby had, indeed, delivered two adorable kids.

Ruby was one of several does that we bred. We opted to breed our own goats to have more control over the quality of the kids in the hopes of producing a better stock of goats for my daughters to show. But there are many other reasons why small-scale goat farmers choose to breed their own goats on their hobby farms.

Like us, other hobby farmers might be interested in maintaining or improving certain traits. When you carefully select a breeding pair, you can develop a line of goats that excel in that desired trait. When you acquire young kids from an outside source or breed one of your does with an outside buck, you don't have this same level of control over the genetic makeup of the offspring.

Hobby farmers often embrace sustainable practices, and breeding their goats aligns with this goal. It means you don't have to rely on purchased goats from outside sources. You can keep your herd population consistent—or expand it if you want—all within the confines of your hobby farm. Some hobby goat farmers are passionate about specific breeds of goats, particularly when it comes to uncommon or heritage breeds. By breeding these goats, these farmers can ensure the responsible protection of this breed, maintain genetic diversity, and prevent the breed from becoming endangered.

Female goats only produce milk after they have given birth. As with all mammals, this is nature's way of providing a ready food source for

Photo Courtesy of Caroline Huntley Pondicherry Farms

FUN FACT
Windows to the Soul

If you've ever looked a goat in the eye, you will have noticed that its pupils are distinctly different from a human's. Goats have horizontal, rectangular pupils that have intrigued farmers and scientists for centuries. These unique eyes are believed to be an advantage to herbivores. While vertical pupils help with depth perception, making it easier to judge how far to pounce, horizontal pupils yield a better field of vision, allowing prey to spot predators and move to safety. In addition to their unique pupils, goats can rotate their eyes in excess of 50 degrees, enabling them to keep their pupil slits parallel to the ground even while grazing.

the young. Most hobby farmers are interested in keeping goats for the milk they provide. After they deliver their kids, does will continue to produce milk for roughly two years as long as they are milked on a consistent basis. In order to keep a steady supply of milk, the does will need to be bred again after their milk supply dries up.

If you are using your hobby farm as a way to educate your children about responsible farming practices, biology, and genetics (and you should), the breeding and birthing process is full of teachable moments. You can discuss genetics and animal husbandry as you select a breeding pair, reproduction during the mating process, and the miracle of birth as you watch the doe kid. There are additional lessons, as well, after the baby goats arrive. The newborns will need to be kept warm, bottle-fed, and protected from harm.

Lastly, some hobby farmers breed their goats to give them an income stream. Surplus baby goats can be sold to other hobby farmers, 4-H youngsters, and for meat.

If you are considering breeding your goats for any of these or other reasons, this chapter is for you. We will discuss the reproduction cycle of goats, how to select suitable breeding pairs, responsible mating practices, as well as the kidding process and newborn care.

Goat Twins and Triplets

You might be considering breeding one of your goats to replenish your herd or replace a goat that has died. For small-scale hobby goat farmers with limited space, it is important to know that goats typically have multiple offspring per pregnancy. Goats don't deliver a single kid very often. Twins and triplets are much more common. It is not out of the question for a doe to deliver four kids in a single litter or "kid crop." Keep this in mind if you are limited to the number of goats you can have in your herd. You may have to sell one or more of the kids if your doe delivers more babies than you anticipated.

Photo Courtesy of Jennifer Estabrook

Understanding the Reproductive Cycle of Goats

Mature female goats, like all mammals, have heat cycles during which they are fertile and will become pregnant if mated. Some breeds of goats become sexually mature as young as two months old, though most does have their first heat when they are four or five months old. The length of the heat cycle varies depending on the breed and the individual goat but typically occurs every 21 days and lasts a few days. Does do not, however, experience heat cycles all year long. They are seasonal breeders, and their heat cycles are triggered by decreased sunlight. Mating usually occurs in the fall and winter, so the kids are born in the spring. There are ways for goat breeders to "cheat the system" and breed their goats at different times of the year if, for whatever reason, they want kids born in the summer, fall, or winter.

Since does are only fertile for a few days, it is important for goat farmers to be able to recognize the signs of heat. It helps if you are well aware of your does' normal behavior and appearance so you will be able to notice when there are changes. A doe in heat will act differently. Some of them will become more friendly and will crave your attention when they are in heat, while others may prefer to be alone.

One common sign of being in heat is tail flagging. Does will hold their tails high in the air and wag them often. It is their way of, quite literally, waving a flag to announce their fertility to all the bucks in the area. Increased vocalization is another common sign of heat. The female goats will bleat and baa more often than normal. Once again, this is an involuntary reaction that is nature's way for does in the wild to attract a mate. Both tail flagging and vocalizations will drastically increase if your does come in close contact with a buck. I know some goat farmers who will lead a buck through the goat barn and note which does react like lovesick schoolgirls and which ones aren't that interested.

In addition to behavioral changes, you will also notice some physical changes when a doe is in heat. You may observe, for example, some vaginal discharge and swelling on the animal's vulva. When a doe goes into heat, you will also notice a decline in her milk production.

How Can You Cheat the System?

As mentioned, goats are seasonal breeders. They will naturally go into heat in the fall and winter in time to have their kids in the spring. Just because that is the natural reproductive cycle of goats doesn't mean it can't be tweaked to accommodate your time frame. There are a few ways you can force a female doe to go into heat.

The first method involves recreating the right conditions to mimic nature. The heat cycle in goats is triggered by the cooler temperatures and decreased sunlight of autumn and winter. If you keep a doe confined in a temperature-controlled stall without natural sunlight, you can adjust the hours of light that she gets. These changes to her environment may trick her body into thinking that winter is coming and she needs to prepare for breeding season.

Goats are also opportunistic breeders and can experience spontaneous heat if put in close contact with a virile buck. The scent of the buck, while offensive to us, is so exciting and intoxicating to does that it is often enough to force them into heat. It is not necessarily the buck; it is the scent. Many hobby farmers will use a "buck rag" to get the job done. It is essentially a rag or piece of cloth that carries the buck's scent. You can rub the rag on the buck or wrap it around the buck's favorite scratching post until it soaks up the powerful odor. Then, tie it around the doe's neck or place it in her stall.

Lastly, hobby farmers could discuss off-season breeding with their veterinarians. A fairly new method of triggering does to go into heat is CIDR, a progesterone suppository that is inserted into a doe. The release of hormones causes the animal to go into heat so she can be bred.

Selecting Suitable Breeding Stock

Your primary goal in breeding your goats is to ensure the overall quality, health, and productivity of the offspring. To accomplish this, it is important that you select the right goats to breed with each other. Seek out the advice of experienced goat breeders or discuss responsible breeding practices with your veterinarian. When selecting a breeding pair, here are a few things to keep in mind:

Breed

Part of responsible breeding is maintaining the integrity of the goat breed. Therefore, you should only breed together goats of the same breed. While "mutt" or mixed-breed goats may produce a lot of milk or be great companions, they are diluting their breed.

Pedigree

If you are planning to enter your goats in goat shows, sell the kids for top dollar, or produce high-quality animals, you will want to select breeding pairs based on their pedigree and bloodlines. From the animals' pedigrees, you can determine the best genetics to influence such traits as milk production, health, and disease resistance.

Temperament

For hobby farm goats, a calm, friendly, nonaggressive temperament is desired. Easy-going, docile goats are more manageable and more family-friendly. The temperament of goats can be passed on to their offspring, so you should look at personality, attitudes, and temperament when matching breeding pairs.

Overall Health

Both the doe and the buck should be in good health, free of disease, and without any major health concerns prior to breeding. Look for goats with robust appearances, that are well-fed but not overweight, and which appear to have good vitality. Goats with chronic or ongoing health conditions should not be bred at all.

Traits

Determine the traits you most desire in the offspring and the purpose of the goats. Knowing this information can help you select breeding pairs that can accomplish your goals. If milk production, for example, is important to you, you should look for a breeding pair with a reputation for producing great milkers.

Age

Even though most female goats are sexually mature by the time they are six months old, most experts strongly suggest that hobby farm goat breeders hold off breeding their does until they are at least one year old. Pregnancy and birth will be easier on the goats' bodies if they are fully grown. Bucks, on the other hand, can be younger. At least seven months old is ideal. Unlike humans, female goats do not go through menopause. Does will remain viable throughout their life span, so theoretically, does can be bred even after they have entered their golden years. But you should be aware that pregnancy and kidding-related complications are much higher in older does and often result in death. Many hobby farmers choose to retire their does from breeding when they are about nine years old to give them a break in their old age.

How to Find a Breeding Buck

Most small-scale goat farmers prefer not to keep a buck on their farms because bucks are smelly, aggressive, and require separate housing. At breeding time, these farmers must find a suitable buck to mate with their doe or does. How do you find such a buck? There are a few ways.

Word of Mouth

If you know other members of the goat community in your area, ask around. Chances are there is someone who would be willing, for a stud fee, to mate their buck with your doe. Just be sure you follow the best practices listed above when selecting a buck to make sure you are being responsible in your breeding practices and that you are doing all you can to produce good-quality kids.

Photo Courtesy of Eva Jordan

Goat Breeder

Some hobby farmers and most larger goat farmers keep a few bucks on hand specifically for breeding and generate a second-ary income stream from stud fees.

These types of farms generally have a larger selection of bucks, maintain detailed breeding records, and have a good track record of responsible breeding. They have the expertise to help you, as well. They can offer suggestions and guidance to help hobby farmers who are new to the goat breeding endeavor.

Ask Your Vet

Talk to your veterinarian about your plans to breed your goats and your search for a suitable buck. He or she may be able to suggest a hobby farmer near you or point you in the direction of resources that can help you find a mate for your doe.

Social Media Groups or Marketplaces

You may be able to connect with a goat farmer in your area who has a buck by posting in Facebook groups for hobby farmers in your town or by responding to posts by other people. Just be sure to fully vet goat owners you meet online, inspect their farm and their animals, and ask for health records and the reproductive history of their bucks to be sure they take the care and health of their animals seriously.

Proper Mating Techniques and Pregnancy

Every time we bred our goats, we made arrangements for stud services with a goat farmer we connected with through our 4-H group. We brought our doe to their farm and introduced her to the buck. The breeding pair was put together in a stall, and from there, they figured out what to do. They didn't need any directions or assistance from us.

If the doe isn't already in heat, being in close proximity to a buck will certainly cause her to go into heat soon. The farmer watched the doe for signs of heat and observed the pair mating.

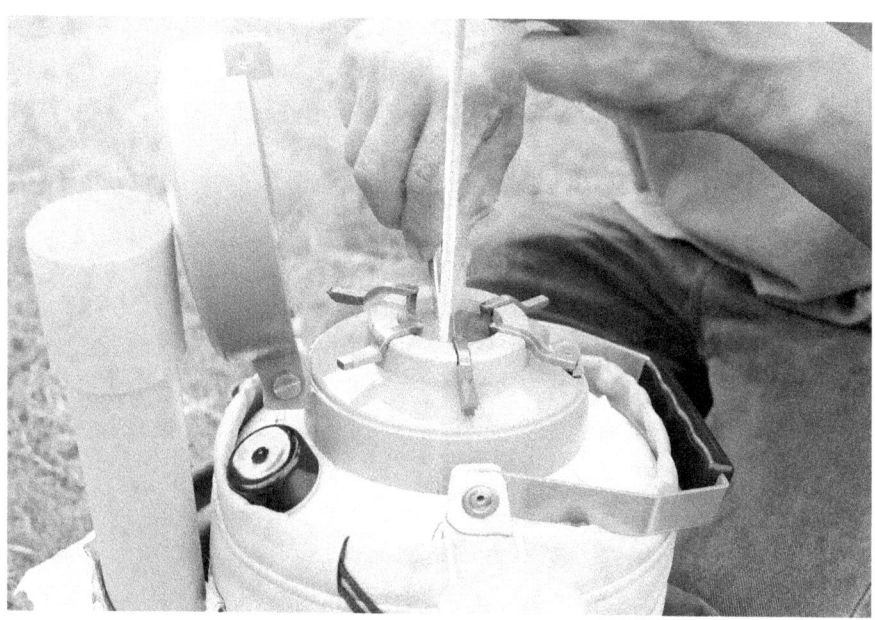

Alternatively, the doe can be artificially inseminated by manually inserting semen into her with a bulb syringe. Although this method does not need to be done by a licensed veterinarian, you may want to discuss the procedure with your veterinarian, especially if it is your first time doing it.

Twenty-one days after her visit with the buck, watch the doe to see if she goes into heat again. If she does not go into heat at her next scheduled time, it is very likely because she is pregnant. Approximately two weeks after she has mated, there will be signs that she is pregnant. One such sign is belly tightening. If you press your fingers in front of her udders, you will feel a tight, tense belly if she is pregnant and a soft, squishy belly if she is not.

Within a month or two of being bred, you will notice that the doe's sides or barrel expands. As her pregnancy progresses, you may observe that the right side of the doe protrudes out further than the left side. In older does, especially ones that have been bred many times, these signs will not show up until four to six weeks before she delivers.

Assisting with Kidding and Caring for Newborn Kids

Gestational Period

The gestation period of a goat is, on average, 150 days. Since you know the date of mating—at least within a few days or a week—you can count out 150 days on your calendar to get an approximate due date. Remember that this is just an approximation. The doe can give birth a few days before or after this date.

Pregnancy Nutrition

Proper nutrition is essential during pregnancy. Be sure to feed your pregnant doe a balanced diet of high-quality hay. A 50/50 mix of alfalfa and hay is ideal. In addition, offer grain and mineral supplements to the goat. A doe who is carrying multiples will need more nutrition than a doe carrying a single kid, but since you will not know how many kids she is carrying, continue to feed her the best food you can give her. Water is equally important. Pregnant goats should have access to fresh, clean drinking water at all times.

Preparing for Kidding Day

To ensure a smooth and successful delivery, make preparations ahead of time for your goat to give birth. This will ensure that you are ready when the big day arrives, and you won't be scrambling to put things in order as your goat labors.

As the date on the calendar grows closer, separate the pregnant doe from the rest of your herd. Provide her with a clean, quiet, private stall, so she has a place where she can give birth without the stress and noise of the other goats. Lay down a thick layer of clean bedding in the birthing

stall. Clean and replace the bedding each day leading up to her delivery—and after. This will reduce the risk of the newborn kids contracting an illness or becoming injured.

Put together a "kidding kit" with the supplies you might need during the delivery process. It should include clean towels, gloves, scissors, lubricant, and iodine for dabbing on the umbilical cord. If the weather is cold, you will want to have a warming box or heat lamp ready to use.

Familiarize yourself with the normal kidding process so you know what to expect and how to identify potential problems. Goats have been giving birth, both in the wild and on farms, for thousands of years without interference from humans, so you should not try to assist during delivery unless you are certain your doe is having some type of complication. By understanding the stages of labor and delivery, you will know when to step in to help and when to let nature run its course.

Signs of Labor

When a doe is in labor, there are several noticeable signs and behavioral changes that indicate the impending birth of her kids. It is essential to closely monitor your pregnant goat during this time to ensure a smooth delivery and the safety of the kids. How do you know when your goat is in labor? She might act restless and agitated. You might observe her pawing at the ground and rearranging the bedding—typical nesting behavior. She might be more vocal or separate herself from the rest of the herd.

There will also be physical signs. The tail ligament, located on each side of the base of the tail, becomes softer and more relaxed as labor begins. You will also observe vaginal discharge that will clue you into the onset of labor. Likewise, you may see rhythmic contractions on the goat's sides and abdomen. You may even see or feel the kids moving inside the doe's belly, especially during the late stages of labor. In some cases, the doe's water may break, and you will see a discharge of clear fluid.

Stages of Labor and Delivery

There are three stages in the labor and delivery process in goats, each with its own characteristics, milestones, and time frames. The first stage focuses on cervical dilation and can last up to 24 hours. During this

stage, the doe may appear restless, show signs of discomfort, and be more vocal than normal.

The active labor stage is the second stage, but it is the stage that results in the birth of the kids. Active labor typically lasts from two to four hours. During this time, the doe experiences contractions and may strain to push. The amniotic sac containing the kid or kids is expelled, and you will see the kid's feet and nose present first.

After the kids are born, the doe will deliver the placenta. This usually happens within a few hours, but it can take up to 12 hours. The new mother will likely be exhausted but will tend to her newborns. She will lick them clean and nurse them. When the placenta is expelled, make sure that the entire organ has passed. If a portion of the placenta is retained, it can cause health issues for the doe.

Record Keeping

Keep detailed records of the kidding process, including the date and time, number of kids, and any complications. Note the size and weight of the kids, their gender, and colorations. Record-keeping is part of good goat herd management on your hobby farm.

To Bottle-Feed or Not to Bottle-Feed

Before the newborn kids arrive, you have a choice to make. Will you allow the mother to naturally nurse the kids, or will you bottle-feed the babies? Or both? There are pros and cons to each approach.

The main reason why hobby goat farmers choose to bottle-raise their kids is so they can keep the mother goat's milk for themselves. For small-scale farmers who are raising dairy goats, milk is a valuable commodity. It is more cost-effective to feed the kids a milk replacer. There are other benefits as well. Bottle-fed babies are friendlier and easier to handle because they bond with their human caregivers. It also makes it easier to wean the kids. Some diseases, like CAE, can pass from mother to kids through milk, so bottle-feeding can reduce the spread of disease.

On the other hand, bottle-feeding newborn kids can be labor intensive. They will need to be fed twice a day on a consistent schedule for two or three months. You need to make sure that you can fit bottle-feeding into your schedule. You will also need more equipment, including bottles and nipples. These will need to be cleaned and sterilized after every feeding.

When you allow the mother goat to nurse her offspring, you will be using a natural feeding method. For many hobby farmers, the natural nursing method aligns with their sustainability goals. Admittedly, it is less work for you as a farmer to allow the mother goat to feed her babies. It is also less costly since you will not need to purchase milk replacer, bottles, and nipples.

One of the drawbacks of nursing, in addition to less milk for you, is that you cannot monitor how much milk each kid is consuming. Nursing can also cause uneven udders if the kids favor one teat over another. Kids who are nursed do not develop close bonds with humans as youngsters and will be harder to handle as they grow older.

Many hobby goat farmers opt to use a hybrid approach to feeding newborn kids. They allow the babies to nurse from their mother for the first few weeks of their lives. After this time, the farmer separates the kids from their mothers and introduces them to bottle-feeding. One of the biggest benefits of this method is that the kids consume colostrum in their mother's milk. Colostrum is the first milk produced by all female mammals, including goats, in the early days after giving birth. It is a highly nutritious and beneficial milk that is designed by nature to give newborns a boost of immunity.

Switching baby goats from the udder to the bottle might be challenging at first, especially because the kids are not used to being handled. It may take a bit of work, but the kids eventually catch on.

The decision whether to bottle feed or mother feed your kids should be made based on your personal preferences and your goals for your hobby farm. We bottle-fed our kids because we wanted to use the milk for ourselves. As I type that, it sounds selfish, but I can assure you that the kids were well-fed and well cared for. My daughters stepped up and helped with the bottle-feeding chore, although the cleaning, sterilization, and formula mixing fell to me. There is no right or wrong way. It all comes down to your situation.

Summary

For small-scale hobby farmers, breeding their own goats can be a rewarding experience. It offers a way to expand your herd, take control of the genetic makeup of your goats, maintain sustainability, and provide an educational event for your family. With an understanding of the reproductive cycle of goats, you can determine when to breed your goats, whether you have a buck on your farm or you have to find a suitable stud. A pregnant goat requires some special care, especially as her delivery date draws near. As a hobby goat farmer, you will need to prepare a birthing stall and, if necessary, assist in the birthing process. Once the kids arrive, you will need to determine whether you will allow the mother to nurse her offspring or bottle-feed the babies.

CHAPTER 7

Goat Nutrition and Feeding

Goats have a reputation for being garbage disposals. Although they might eat everything in sight—including my newly planted pear trees—that doesn't mean they should. Proper nutrition is essential for goats because it directly impacts their overall health, well-being, and productivity. Not only does a balanced diet help the animals fight off diseases and maintain body functions, but it also directly impacts the quality of the milk, meat, and fibers that they produce. You can't expect to get maximum output from your goats if they are not given the best fuel.

People who have never raised goats before may assume you can just release a herd of goats into a pasture and let them feed themselves. There is more to it than that. Goats are ruminant animals with unique digestive systems, as we will see momentarily. They can digest tough, fibrous plant material that we humans cannot digest. Their ruminant system also means that goats need a combination of foraged food, roughage, and concentrated feed to meet their nutritional requirements.

In this chapter, we will discuss exactly what those nutritional requirements are, how they change throughout a goat's lifetime, and the best diet for different types of goats—dairy, meat, or textile. We will take a look at pasture management practices and list some of the plants that are toxic to goats. Lastly, we will go over how to decode the nutrition label on commercial goat feed and how to know if you need to offer mineral supplements to your goats. When you provide a well-balanced and appropriate diet for your hobby farm goat herd, you will be taking an important step toward ensuring the overall health of your goats and contributing to the success of your goat farming endeavors.

The Goat's Four Stomachs

Goats, along with cows, sheep, and deer, have specialized stomachs with four separate compartments. This unique digestive system allows them to efficiently digest the fibrous plant material that is essential for their diets. The four chambers are the rumen, reticulum, omasum, and abomasum. Each compartment has its own specific function in the digestive process.

The rumen is the largest compartment in the animal's stomach, and it functions much like a fermentation vat. Within the rumen are billions of beneficial microorganisms, including bacteria, fungi, and protozoa. These microorganisms go to work producing enzymes to break down the fibrous plant material that the goat eats, including the tough cellulose that other animals cannot digest. The plant material is deconstructed into simpler compounds, such as fatty acids and gases, including methane and carbon dioxide. These by-products of the fermentation process are then absorbed through the walls of the rumen to give the goat a significant

RUMINANT STOMACH

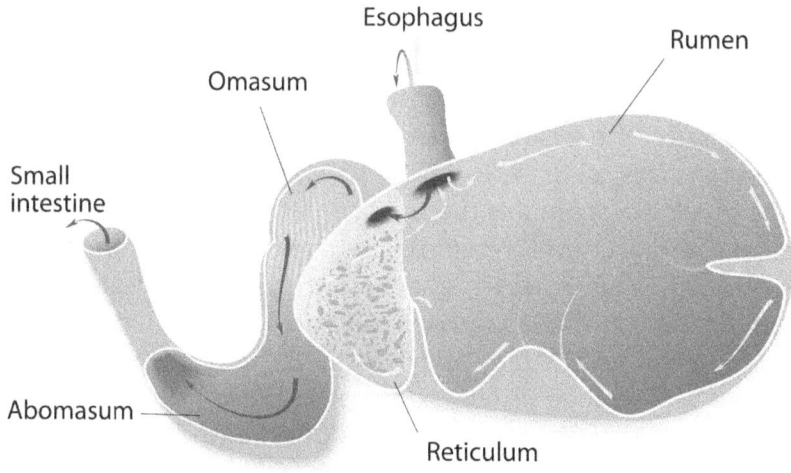

Most goats enjoy a varied diet of nutrient-rich grasses and feed. However, when given a chance (or job) to munch on weeds and shrubbery, many goats won't miss the opportunity. Goats are equipped with an impressive set of teeth designed to masticate tough stems and vegetation. Like humans, goats go through two sets of teeth in their lives. A goat's baby teeth are replaced by the first set of adult teeth between one and two years old, and most goats have a full set of adult teeth by age four. Goat teeth comprise molars and incisors, but they have no canines or top front teeth. Instead, goats have a "dental pad" at the front top of their mouths. This dental pad functions like a top set of teeth, but most of a goat's chewing happens with its molars.

source of energy and nutrients. But this is only the start of the digestive process.

The reticulum is located just behind the rumen. It acts as a storage compartment for the remaining coarse plant materials. It is sometimes referred to as the honeycomb because its lining is grooved and ridged like a honeycomb. These grooves and ridges collect the larger, indigestible items that the goat may have eaten, like stones, hard plastic, or the iconic tin can that goats are reported to love.

The remaining food particles then pass to the omasum, a chamber that contains many folds. These folds essentially squeeze the water and fluid from the partially digested food so more of the nutrients can be absorbed. This process further breaks down the size of the remaining food, turning it into tiny particles before it is passed to the last of the four chambers, the abomasum.

The goat's abomasum is the true stomach of the animal. It is the chamber that most closely resembles the stomach of non-ruminant animals. The abomasum secretes digestive enzymes and hydrochloric acid to further break down the food particles into simpler components, like sugars, amino acids, and fatty acids. Finally, the food particles are small enough to be absorbed through the goat's stomach lining and into its bloodstream. Although the abomasum does the work of a traditional stomach, it would not be able to do its job without the previous three stomachs. All four of the stomach compartments work together to efficiently utilize the plant nutrients.

Do Goats Really Eat Everything?

As I mentioned earlier, goats have a long-held reputation for being non-picky eaters. Nearly every cartoon goat you see is shown chewing on a tin can. This reputation must have come from somewhere, but is there a kernel of truth in it? Not entirely. While goats are indeed opportunistic feeders with a relatively broad diet and a habit of sampling new things, their ability to eat everything is often exaggerated and misunderstood.

One reason for this misconception could be that goats in the wild thrive in rugged, rocky landscapes where other animals may struggle to find sufficient food to eat. Goats are known for eating vegetation that other animals cannot or will not eat. Perhaps the source of this reputation stems from the curiosity of goats. Goats like to explore their surroundings, and their investigations often include nibbling on various objects. It is likely that farmers of the past observed goats with random items in their mouths, like a tin can.

The reality is that goats are not indiscriminate eaters. Although they do have a broad diet and are adaptable to new environments, they are quite selective in their browsing and grazing. They prefer some foods over others but will eat foods they don't like if their favorite items are not available.

Types of Goat Food

Even though goats' diets are broad, they can be separated into three categories: goat feed, hay, and pasture grazing. Goats' digestive systems work best when the animals are fed a combination of these three. It also ensures that the goats are getting all the vitamins, minerals, and nutrients they need to grow strong and stave off diseases. Let's take a look at each of these categories.

Goat Feed

In the wild, goats do not eat large quantities of grain. They prefer plant material, like leaves, grasses, and small tree branches. Yet, as goats became domesticated, their diets evolved. Goats confined to pens or pastures were not free to browse the land in search of the food items they craved. They could only eat what grew in their pastures or what the farmer fed them. Feeding goats grain—usually a mix of corn, soybeans, and oats that had been shredded and formed into pellets or mixed in a loose blend—was a necessary step to ensure the animals were getting the nutrients they needed.

Commercial goat feeds that you can buy at your local farm supply store are all subject to a strict set of regulations to make sure they meet or exceed the standard that has been established by the Food and Drug Administration. The FDA works with experts in livestock nutrition to continuously tweak these guidelines as new information becomes available. Commercial goat feed manufacturers are also required to print accurate content and nutrition labels on all their packaging. Later in this chapter, we will discuss how to read nutrition labels.

It is important to point out that sweet feed is not the same thing as goat feed. You will probably have some experienced goat farmers tell you that it is okay—and cheaper—to give your goats sweet feed rather than feed that is specifically labeled as goat feed. Just what is sweet feed, and how is it different from goat feed? The answer is that it is sweet!

Sweet feed is a blend of grains, like corn and soybeans, that have been processed into pellets using molasses as a bonding agent. Molasses is ideal for holding the grains together, and it contains iron, which goats need. But it is also a type of sugar. Think of it as the sweet, sugary coating on breakfast cereals that are marketed to children. Sure, there is whole grain involved, but the entire thing is drenched in sugar. It is no wonder that kids—both goat and human—love it.

Goats can easily fall in love with sweet feed. Once they get a taste of sweet feed, they will turn their snouts up at goat feed that isn't made with molasses. Just as in humans, however, too much sugar can be detrimental to goats' health. Sweet feed should be used as a treat and given sparingly, if at all. Stick to non-sweet commercial goat feed as the staple of their diet.

Many small-scale hobby farmers, especially those who value sustainability, choose to make their own goat feed using organic grains they grow themselves. This is admirable and falls in line with the goal of being self-sustaining. As with most things, there are benefits and drawbacks to making your own goat feed. The benefits are that you will have more control over the ingredients that go into the feed. You have the ability to make sure the grains have not been genetically modified and are free of chemical pesticides. If you grow the grains yourself, you could potentially save a lot of money using homemade feed rather than buying commercial goat feed at your local farm and feed store.

However, you could find that it is more costly to mix your own goat feed, which is a drawback. You will have to grow a wide range of grains and vegetables—and grow enough of them to sustain your goat herd until the next harvest. You will need to be able to properly dry and store the grains without them spoiling, which may require a sizable storage room or building. And then there is the time commitment that is required for you to mix your own goat feed. Lastly, there is the possibility that your homemade, organic goat feed will be lacking in one or more essential nutrients, even if you faithfully follow one of the many organic goat feed recipes available online. Without laboratory equipment, you cannot detect the exact minerals in your batch of feed.

Hay

Hay is basically the grasses, plants, and legumes that your goats eat fresh in their pasture, only they have been cut and dried to make them last longer. Hay is made by cutting mature grasses and plants that are at their peak of nutritional value before they start producing seeds. The cut plants are thoroughly dried to remove all the moisture. Hay is commonly used as a primary food source for goats and other livestock animals, especially over the winter months when access to fresh vegetation is limited.

Hay provides goats with a source of roughage that is crucial for their digestive health. Without a diet of roughage, a goat's ruminant system will fail to function properly. Hay is high in fiber, stimulating the animal's

*Photo Courtesy of
Gayle Ewen*

rumen to ferment the food. Although hay is not as nutritionally rich as fresh green vegetation, it does contain a variety of important nutrients, including carbohydrates, fats, proteins, vitamins, and minerals.

You will encounter goat farmers who only feed their goats hay in the winter months, as well as farmers who offer hay to their goat herd year-round. In the winter, hay becomes a primary food source for goats, but they can benefit from eating hay during the summer months, too. In some situations, especially on small hobby farms, the pastures become overgrazed, and the forage is limited. Year-round hay is a good way to supplement your herd's diet. It is crucial that you provide your goats with good-quality hay that is free from mold, dust, or toxic plants.

An alternative to traditional hay for goats is alfalfa pellets. Alfalfa pellets are made by shredding and pulverizing alfalfa hay into alfalfa meal, which is then dried and pressed into pellets. A benefit of feeding your goats alfalfa pellets is that there is little waste involved. If you have watched your goats eat hay from their manger, you have probably noticed that a good quantity of the hay ends up on the ground, gets trampled on, and goes to waste. The pellets are easier for the goats to eat without ruining a portion of their meal. Alfalfa pellets are packed with protein, vitamins, and minerals; however, they do not contain the roughage that your goats need. Goats still need hay and/or grazed vegetation in their diets.

Photo Courtesy of
Mary Beth O'Donovan

Forage

As we have noted, the digestive system of goats is uniquely designed to digest fresh, tough, fibrous plant material. It is part of a goat's makeup. Another part is their foraging instinct. Goats in the wild spend their days wandering from place to place, snacking on tasty vegetation. Domestic goats still have a strong desire to graze or browse. It is an important survival tactic and an ingrained, natural behavior. The act of grazing and browsing, in fact, stimulates the goat's rumen to produce enzymes to break down the food.

Grazing and foraging also offer a good form of exercise for goats, which helps them maintain muscle tone and prevents obesity. Curious and intelligent animals, goats find mental stimulation from foraging and browsing. It keeps them mentally engaged and prevents boredom. Goats are also herd animals that crave companionship. Grazing in a herd setting allows goats to interact and establish their social hierarchy. The key, as we will see next, is to have a well-maintained pasture with a variety of vegetation.

Pasture Management

Since grazing for vegetation is a significant food source for goats, it is essential to manage their grazing habitat responsibility. Overgrazing can diminish the quality of vegetation, cause soil depletion, and make regrowing desirable plants difficult. When you first establish your small-scale goat farm and fence off a pasture area, your first order of business is to rid the pasture of noxious plants and trees that can sicken your goats.

Several trees are poisonous to goats. If ingested, they can cause various health issues, even death. Wild cherry and black cherry trees, for example, contain a substance in the leaves, bark, and branches that metabolizes as cyanide when eaten. Likewise, black walnut trees produce a toxic substance that can sicken goats. All parts of yew plants are highly toxic to goats and can lead to sudden death. Goats can experience both gastrointestinal and neurological issues after eating mountain laurel leaves or branches.

Among the bushes and shrubs that are toxic to goats are rhododendron and azalea. Boxwood shrubs, forsythia, and holly can also be harmful to goats if ingested. As for plants, nightshade, pokeweed, lily of the valley, hemlock, buttercups, and bracken ferns can cause severe health issues if consumed by goats.

Before you turn your goats out to pasture, inspect the plants growing in the space and remove all of them that can be detrimental to the health of your goat herd. This is not a one-and-done task. Every year, you should inspect the grazing area and remove any toxic plants that may have taken root there to keep your goats from sampling them.

In addition to keeping bad plants out of your pasture, you want to encourage healthy, beneficial plants to grow there. A well-maintained pasture will have a diverse range of safe and nutritious vegetation that meets the dietary requirements of your hobby farm goats. Since overgrazing can be a problem, especially in small pastures, set up a system to rotate your pasture areas. Move your goat herd to a different grazing area periodically to allow the vegetation in the first pasture to recover and regrow.

Consider reseeding your pasture to improve the quantity and quality of the forage available to your goats. In fact, reseeding is one of the best practices for maintaining a healthy grazing area. Start by selecting the right grasses and plants for your region and soil type. Choose ones with high nutritional value for goats. Next, determine the ideal time to reseed the pasture. This will depend on your climate and weather conditions, but generally, reseeding takes place in early spring or late fall. Prepare the seedbed by tilling the land or aerating the soil to break up compacted areas and allow the seeds to reach the soil. Spread the seeds over the area in a uniform pattern, lightly cover the seeds, and keep the soil moist. Keep your goat herd out of this pasture area until the new plants and grasses grow tall enough and are firmly established.

Just like monitoring your pasture for noxious plants, reseeding your pasture is not a one-time activity. It is a task that should be done periodically to maintain the productivity and robust health of the pasture. Part of proper pasture maintenance is employing a combined plan of pasture rotating and reseeding.

Nutritional Requirements for Different Stages of Goat Life

The nutritional requirements of goats vary for each stage of their lives, from birth through their golden years. We know that goats need to eat a diet that is high in fiber and roughage, as well as one that contains the proper mix of vitamins and minerals. Calcium, iron, protein, potassium, phosphorous, magnesium, iodine, copper, and zinc are all key nutritional requirements.

Newborn Goats

Colostrum is the first milk produced by goat mothers, as well as all female mammals, including cows, horses, sheep, and even humans. It is produced shortly after the female gives birth. Colostrum differs from regular milk in that it is thick, rich, and yellowish in color. In fact, its color—and its vital importance to newborns—has led to its nickname "mother's gold." Colostrum is packed with nutrients and high in antibiotics to give newborn kids an immunity boost. Unfortunately, colostrum is a fleeting substance. Does only produce colostrum for a short time, typically about

24 hours. For this reason, newborn kids should drink as much colostrum as possible shortly after birth.

Many hobby goat farmers are breeding goats so they can acquire their milk, so they bottle feed the newborn kids using a milk replacer formula. While bottle feeding allows you to more closely monitor the amount of food your kids eat, be sure to feed the correct amount of the milk replacer. Avoid overfeeding, as this can be detrimental to young kids. Starting on their second day of life—after allowing the babies to dine on colostrum their first day—bottle feed your goats four to six ounces of milk replacer four to five times per day until the kids are 10 days old. After that, from age 11 days to 21 days, offer the kids bigger portions—7 to 12 ounces—three to four times per day. When the goats reach 21 days old, they should be drinking 12 to 16 ounces of milk replacer three times a day until they are weaned.

Young kids will instinctively start to nibble on hay and plants even during the time they are being bottle-fed. Don't try to prevent this. It is good training to prepare them for browsing when they are weaned. Once they are weaned, young goats need a diet with sufficient protein content to ensure their muscle and tissue development. Very young, bottle-fed kids need 14% of their diet to be protein. Yearling kids need 12%. They also need energy from carbohydrates and fats. Proper mineral and vitamin supplements are key for healthy bone development and metabolic function.

Pregnant Does

Pregnant does in the first three months of pregnancy should be fed their regular diet and do not need extra food. If the doe is at her optimal weight already, additional food will just increase her chances of becoming obese, which can make her pregnancy and delivery more difficult. A doe should, instead, drink additional water throughout her pregnancy. A pregnant doe, in fact, can require as much as four gallons of clean, fresh water per day until she delivers. You may need to check her water supply several times throughout the day to make sure she doesn't run out and adjust your watering schedule to plan for more frequent refills.

The doe's dietary needs change during the last two months of pregnancy because more than 70% of kid growth takes place during this time. That means the doe's nutritional and energy requirements greatly increase. She will need to increase her protein intake from about 6% to about 12.8%. Her total digestive nutrient (TDN) requirements—the measure of energy from the fiber, carbohydrates, vitamins, and other minerals in feed—will also increase, going from about 52% to 66%. The amount of dry food, such as hay, should also be increased from about 2.8 lbs. to around 4 lbs. It is difficult for a pregnant goat to consume this amount from grazing alone. She should be fed supplemental grain and hay.

Inadequate nutrition during the last part of a doe's pregnancy can have major consequences on her health and the health of her unborn kids. Does can suffer from pregnancy toxemia, a condition that occurs when the goat's pregnant body requires more nutrients than she is consuming. To compensate, nutrients are pulled from her body's reserves, weakening her and putting her at increased risk for disease. The in-utero growth of her kids is also impacted. Feeding pregnant does a proper diet of high-energy, high-protein food can combat this.

Adult Dairy Goats

Adult lactating dairy goats need a well-balanced diet to support their milk production, as well as their overall health. Like all goats, lactating dairy goats need the fiber and roughage found in hay and foraged foods to keep their ruminant system functioning properly. They also require a

balanced supply of essential minerals and vitamins, including calcium, phosphorus, potassium, magnesium, sodium, zinc, and vitamins A, D, and E. Lactating goats need an increase in protein and energy in their diets to support milk production. Depending on how much milk they produce per day, their protein requirements can be as much as 20%, and their TDN can reach 11.6%.

Inadequate nutrient intake can greatly impact the quantity and quality of the milk the doe produces. If her dietary needs are not met, she might also stop producing milk altogether. Hobby farmers who are looking to earn an income from their dairy goats need to be willing to invest in quality, nutritious food to keep their herd of dairy goats productive and healthy.

Meat Goats

For meat goats, proper nutrition to build muscle tone should start when the kids are very young and continue until they go to market. The more attention devoted to ensuring your meat goats are fed an adequate amount of a nutritious diet, the better quality the meat will be. An adult buck, for example, should consume a diet with 11% protein and 60% TDN. An animal that weighs between 80 and 120 pounds should eat about five pounds of dry matter each day. Resist the urge to overfeed meat goats to fatten them up before market. The best meat comes from toned, muscular, well-built goats, not from overweight or obese ones.

Understand the Nutrition Label

There is a lot of information to be gleaned from reading the nutritional labels on commercial goat food. The labels tell you more than just the ingredients in the product. They also inform you about the intended use of the feed. For example, you will find some sacks of feed that are labeled "meat goat" feed. These have been designed, with the assistance of livestock nutritionists, biologists, and food scientists, to meet the specific needs of meat goats and to aid in building muscles. A dairy goat formula will focus more on giving the goats the nutrient boost they need to produce high amounts of good-quality milk. If you have both dairy goats and meat goats on your hobby farm, it is in your best interest to buy separate feed for each to make sure you are hitting each group's nutritional requirements.

The label will also offer you a recommendation on how much you should feed your goats based on their age and weight. Overfeeding is a common problem for new hobby goat farmers. In their quest to take the best care of their animals, they can sometimes overdo it. Obesity in goats can lead to health consequences, just like it can in people. Use the information on the label of the goat feed as a guide to help you make sure your goats are fed enough—but not too much.

The first item listed on the nutritional label will be the protein content. It will be called crude protein and given a percentage. Yearling kids, pregnant does, and meat goats need higher levels of protein; therefore, you can use this information to help you decide which feed is best for your situation.

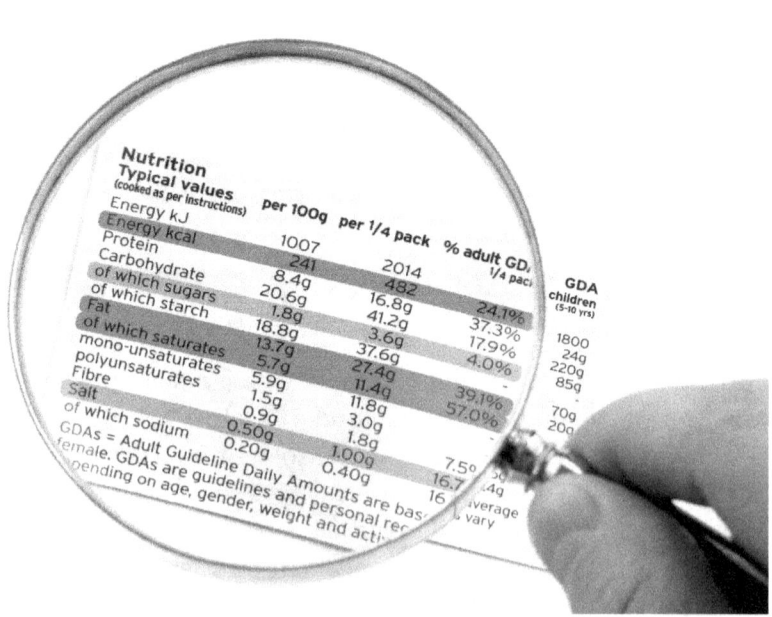

Some labels may have the word "medicated" on them. What this means is that decoquinate has been added to the feed to prevent coccidiosis in the animals. Some hobby farmers prefer to treat illnesses as they arise or would rather have more control over the preventative health of their herd, so it is important to know if the feed has been treated with medication or not.

Feed labels will also list the most common minerals, such as calcium, phosphorus, copper, and vitamin A, and the percentages of each one. If your goats suffer from one of the many illnesses caused by mineral deficiencies, you will want to know how much of the lacking mineral can be obtained from the feed. This will let you know if you should offer mineral supplements to your herd or not.

Should You Offer Mineral Supplements?

Mineral supplements are products that are designed to provide goats with essential minerals and vitamins that may be lacking or insufficient in their regular diet. Supplements aim to balance the nutritional profile of your goats' diet.

While it is important that your goats have an adequate amount of essential minerals, too much of a good thing can be bad. You should weigh several factors before determining if you should provide your herd with mineral supplements. First, look for signs and symptoms of mineral deficiency in your goats. An inadequate amount of minerals may manifest in outward signs of sickness or other health concerns, such as poor milk production, rough coat, lethargy, or slow growth.

Consider your geographic location and the quality of vegetation available to your herd. Some areas have soil that has been depleted of minerals, so the vegetation growing there is inadequate. You can ask your county extension office if it provides forage analysis services to help local farmers assess the quality of their pasture lands. The analysis may determine that your pasture is not sufficient to meet the nutritional needs of your goats and suggest you either reseed your pasture or provide mineral supplements.

You should also have a conversation with your veterinarian about mineral supplements. Based on your veterinarian's examination of your goat, he or she may recommend that you supplement your herd's diet.

Mineral supplements for goats often take the form of mineral blocks, large bricks chock-full of minerals. All you have to do is set the mineral block in a place that is accessible to your goats and protected from the elements. Goats are naturally curious and like to sample things with their tongues. They will find the mineral block in no time at all and start licking it.

Practice Good Food Management and Storage

Feed costs will be the greatest goat-related expense on your hobby farm. Accept this and plan for it so you are not tempted to cut corners. Instead, purchase the best grain-based feed and the highest-quality hay you can find. Your goats deserve it.

To help control feed costs, take steps to reduce waste, and do not overfeed your goats. Goats are not dainty eaters. They will spill their food and knock some of it on the ground. Work on methods to keep waste at a minimum, such as not overfilling the manger with hay or installing a catch tray under the food trough. Eager goats sometimes butt each other out of the way to get to the food, spilling it in the process. If this is a problem on your hobby farm, you can try setting up multiple feed areas or feeding each goat separately.

Goat food that spoils and must be thrown away will also hurt your feed budget. Store bales of hay in a dry place, well protected from the elements. If hay gets wet, it can quickly get moldy. Since most hobby goat farmers will purchase a year's supply of hay in the summer, during hay-making time, it is vital that they have a proper place to store it.

Grain-based goat food contains moisture and can spoil quickly. Rather than buying a supply of goat feed that will last for several months at a time, you may want to consider buying smaller quantities more often. Store the food in a sealed container to prevent it from getting wet and to keep mice out of it.

Feed and water dishes must be thoroughly cleaned and disinfected on a regular basis. I clean ours once a week, but honestly, it should be done twice a week or more.

Summary

The food your goats eat is the fuel that will keep them healthy, active, and productive. Goats have a unique digestive system that allows them to eat tough, fibrous vegetation that other animals cannot digest, but the ruminant system means that goats need to browse, forage, and graze for food in their pasture. They may not get all the nutrients they need from grazing; therefore, grain-based goat feed should also be fed to your herd. The nutritional requirements of goats change over the course of their lifetime, so it is important for you to have an understanding of their needs and meet them.

CHAPTER 8

Troubleshooting and Problem-solving

Raising goats on your hobby farm can be a rewarding and enjoyable experience—but goats can be characters. They are inquisitive, playful, determined, stubborn, and smart. While those can be positive attributes in humans, in goats, they can spell trouble. If you don't believe me, just talk to other hobby farm goat farmers. They can tell you some hilarious stories about goat antics that will have you rolling on the goat barn floor, as well as a few stories that will concern you a bit. That is the nature of raising goats.

In addition to Houdini-like goats that often escaped the pasture and perched on top of our cars, we have had a few trickster goats that pushed our buttons and made life interesting on our hobby farm.

You can take steps to prevent problems from occurring, but expect there to be things that slip through the cracks. Be prepared to do some troubleshooting and problem-solving on the fly to address problems that are unique to your farm.

When hobby goat farmers get together and swap stories, a few common themes emerge—some challenges they all face or have faced at one time or another. This chapter will discuss several of these and will, more importantly, offer some solutions to these issues should they arise on your small-scale goat farm.

Fence Jumpers and Escape Artists

It is a big, wide world out there, and your goats want to see it all. They may have the greenest, lushest pasture and the Taj Mahal of goat barns, and they will still want to leave it behind to go exploring. Most of the time,

they are simply curious, but other times, there might be something that tempts them to make their escape—like the pear sapling I planted near their pasture. I learned rather quickly not to plant anything in sight of the goats that might lure them from their pasture. Of course, that hasn't completely solved the problem.

Smart and crafty, goats will find and exploit any weakness in your fencing and housing. Juju, for example, discovered he could squeeze through a small space between the side of the goat barn and a wooden fence post. He worked at it so much that the fence post became loose and wiggled back and forth by a few inches. Those few inches were all Juju needed to become a free man—er, goat. Callie watched us open and close the metal gate so many times that she understood how the U-shaped latch could flip up to open the gate and flip down to secure it closed. She used her prehensile tongue to lift the latch and stroll out through the gate. Unfortunately, she never learned how to latch it behind her.

We've had plenty of world-class high jumpers that could sail over the pasture fence with ease, including Dixie. We ran barbed wire above the fence, and that prevented all the other goats from leaping over it—except Dixie. She was one of a kind!

As a hobby goat farmer, fence and barn inspections, modifications, maintenance, and replacements will be an ongoing task. Once you fix one escape route, the goats will find another. The best you can do is to invest in well-crafted, solidly built housing and fencing, make sure it is installed correctly, and do routine maintenance.

I'd also suggest that you check on your goats frequently so you can corral the escape artists back in their pasture before they get into too much trouble. We've had many fence jumpers over the years, but thankfully, they have always stayed on our property, and we realized they were out before they could get far. I know some small-scale goat farmers who haven't been so lucky. They have had goats trespass onto the neighbor's property and cause costly damage before they even noticed the animals had escaped.

Destructive Goats

My aunt and uncle's goats ate their barn. Yes, ATE it! In a tremendous display of teamwork, the goats worked together to chew an enormous hole in the side of the barn. Once that section was replaced and repaired, the goats set to work on another wall. Eventually, the entire barn was torn down and replaced.

Goats are so intelligent and inquisitive that they get bored easily. They need something to occupy them and stimulate their minds, and unfortunately, mischief is very stimulating. A destructive goat could be craving mental stimulation. You could try adding some enrichment toys to their pasture or barn to give your goats something fun and nondestructive to occupy their time.

In our pasture, as I mentioned before, there is an old fieldstone foundation of a former outbuilding. The goats love to play "king of the mountain" on it. They climb all over it, headbutt each other, and leap off it. We enjoy watching them play. Some other hobby farmers I know have placed some of those big wooden spools used for cable wires in their pastures to give their goats things to climb on. Elevated walkways, ramps, bridges, and lookout posts are also great additions to a goat pasture.

Goats are, after all, big kids (pun intended). They enjoy the same sort of playground equipment that human kids like. Bring in a small trampoline, a tire swing, and a teeter-totter. You could also try a kiddie wading pool and one of those nylon tunnels used in dog agility shows. The goats will figure out how to play with them. Erecting a goat playground will give your herd so much to do that they will forget about being destructive.

Occasionally, goats nibble and chew on things they aren't supposed to eat because they are lacking in certain minerals in their diets. When you use your garden rake in the heat of the summer, for example, your sweaty hands make the wooden handle salty. If a goat happens upon the rake, it might start licking the handle—then chewing on it. If you suspect that your goats are chewing on random things because they are lacking certain minerals in their diet, get a mineral block for them.

Combating Goat Odors

All barnyard animals produce odors. That's just part of the farm-fresh country air. Goat odors, however, can get out of hand and cause big problems—especially if your neighbors catch wind of the aroma. There are things you can do to mitigate the smells coming from your hobby farm so that you, your family, and your neighbors can all breathe easier.

For starters, don't keep a buck. Intact male goats have scent glands located near their horns and hind legs that produce a pungent secretion of pheromones. The pheromones are used to attract mates and for the bucks to use to mark their territory. The strong odor is the buck's natural way to establish its dominance and to warn other males to keep clear. During mating season, when other intact male goats are around and when the female goats are in heat, a buck's glands will kick into high gear, producing even more of this pungent secretion.

This is probably the main reason why hobby farmers, especially ones on small farms, prefer not to keep a buck in their herd. Removing a buck from your herd, however, won't eliminate all the smells. Urine and manure from does and wethers can also be smelly.

If you find that the smell coming from your goat barn is too unpleasant or stronger than normal, it may be time for a deep cleaning. Regular

cleaning of the goats' living quarters—removing and replacing bedding, hosing out the urine and manure, etc.—is part of your weekly farm chores, but periodically, the entire goat barn should be scrubbed and disinfected from top to bottom.

It is not a bad idea to give your goats baths every now and again. Fecal matter, mud, and dirt can stick to their fur and contribute to their rank smell. As long as the weather isn't too cold, your goats will enjoy getting washed and shampooed. And you will appreciate how wonderful they smell afterward.

Some of the smell from your hobby farm can be attributed to the disposal of manure, which brings us to the next point.

Disposing of Manure and Soiled Bedding

We covered composting soiled bedding and manure disposal in Chapter 3, but it is worth revisiting now because this is a common problem impacting all goat farmers, both large-scale commercial farms and small hobby farms. Even the smallest goat herd produces waste materials, such as manure and soiled bedding. Plan for this inevitability when you set up your goat barn so you are not scrambling to trouble-shoot waste disposal solutions in the midst of getting your goats settled into their new home. Composting the waste material is the most popular and environmentally friendly solution for hobby farmers, particularly ones who strive for sustainability.

Other solutions for this problem include spreading the nutrient-rich manure on fields and pastures, delivering the goat manure for use on agricultural crops, or selling it to earthworm farms.

How to Dispose of a Dead Goat

As much as we don't want to think about it, death is a natural occurrence. You will undoubtedly experience the death of a goat. Amid this upsetting and tragic time, you will have a pressing problem you will need to resolve—how to properly dispose of the animal's carcass. In fact, this is crucial for both biosecurity and environmental reasons. Long before you experience the death of one of your goats, you should have a plan in place to address this problem.

One of the most common and environmentally friendly methods is to bury the goat's body. Check the laws and ordinances in your state or community regarding burying animals on your property. In the state of Michigan, for example, farmers are required by law to bury a dead livestock animal within 24 hours of its death and must bury the animal more than two feet deep. The regulations also spell out the required distance from water sources and how far apart each animal grave must be. Learn the laws in your location in advance so you can plan for this inevitability. Select a burial location that meets the requirements. You may even want to dig a hole before the ground freezes, just in case. I am speaking from experience—it is much easier to dig a large, deep hole when the ground is not solidly frozen.

There are a few other options regarding the disposal of a deceased goat, depending on where you live. If you have access to a large enough incinerator, you could consider cremating the goat's body. This method might not be feasible for all hobby farmers. Some areas have rendering services that will pick up dead animals for proper disposal. Rendering involves processing the carcass to produce useful by-products.

What to Do When You Are on Vacation

When you commit to starting a hobby farm and raising a small-scale goat herd, you are agreeing to take responsibility for the care and keeping of living animals. That's a big responsibility. But sometimes, you need a break. Living on a hobby goat farm doesn't mean your traveling days are over. In fact, a family vacation is always a great way to recharge your batteries. Before you leave on vacation, you need to make arrangements for a qualified, trusted person to take over your goat chores during your absence. Just how do you find this person? That can be problematic.

Start with people you already know—family members, neighbors, coworkers, and others. Someone in this group might love animals and welcome the opportunity to help you. If you have joined a group or organization for hobby goat farmers in your community or a 4-H club, there might be someone in that group you can hire. Another hobby farmer might even know a responsible teenager or college-age young adult who would like to earn some extra money.

If you cannot connect with a potential goat sitter via word of mouth, you can try posting an ad on an online job service app or hanging a flier on the bulletin board of your local feed store. Just be sure to thoroughly vet applicants,

DID YOU KNOW
Goat Yoga

Goat yoga is a fitness sensation that's swept the nation in recent years, appearing in locations from city parks to local hobby farms. Some small-scale farmers have leveraged the popularity of this fitness phenomenon as a lucrative method for bringing more people, and hopefully more business, to their farms. Goat yoga is believed to have started in a small town in Oregon in 2017. According to Guinness World Records, the largest goat yoga class consisted of 501 people and took place in Florida in September 2019. This event was hosted by the Grady Goat Foundation and served as a fundraiser to fight human trafficking.

check their references, and meet with them in person a few times so you can determine how responsible and dependable they are.

Have a plan B in place just in case something goes wrong while you are away and your goat sitter can't fulfill his or her duties. They could have a car breakdown, a family emergency, or get called into work. Ask a trusted neighbor or a cousin if they are willing to help out on a standby basis, just in case. Most likely, they will not be called upon to step in, but it will give you peace of mind to know that if your goat sitter experiences a hitch, your goats won't suffer.

For a number of years, our go-to goat sitter was my niece, Maggie. When we traveled, she would stay at our house so she could take care of the goats, as well as the dog, cat, hamster, fish, chickens, ducks, vegetable garden, and houseplants. But the day arrived when Maggie aged out of the goat-sitting business and got a full-time job in another town. We were back at square one. Then, we happened to connect with a new family down the street. The mom mentioned how much they enjoyed seeing our goats and chickens when they walked past our house and that her animal-lover daughter, MaKayla, in particular, loved to see our goats. I jumped on this tidbit of information. I learned that MaKayla was a high school student and a pretty responsible kid who was looking for some dog-sitting or house-sitting gigs. She was thrilled to add goat sitter to her résumé.

I learned with both Maggie and MaKayla that once you find a good goat sitter you can trust, it is important to hold onto them as long as you can. That would be my advice to you. Pay them well, treat them well, and hopefully, they will remain your loyal and devoted goat sitter for several years.

Behavioral Issues in Your Goat Herd

The Aggressive Goat

In general, goats are docile creatures with an easygoing demeanor and a friendly nature. But from time to time, you may run across a goat with an attitude problem. This is more common among bucks, especially when the does are in heat and there are other bucks around vying for their attention. Handling an aggressive goat in your small-scale hobby farm goat herd can be challenging for you and your family. Your safety, as well as the safety of the other goats, should be your top priorities.

If your aggressive goat is a buck, you may want to consider castrating him to reduce the surging hormones that are causing his territorial and dominance-related behavior. You should also house him in a separate location so he cannot harm the does.

Occasionally, a goat will act out aggressively because its current situation is at odds with its natural instinctive desires. For example, goats

have a strong grazing urge. Goats that do not have access to a pasture and cannot satisfy their urges to graze and browse may respond with aggression. This is also true of lonely goats. Goats are natural pack animals, and they need to interact with other goats. A single goat living in a herd of one will be extremely lonely and depressed. Those emotions may manifest into aggressiveness. Do what you can so that your hobby farm meets your goats' needs.

An aggressive goat can disrupt the overall harmony of your hobby farm goat herd. If you are experiencing issues with an aggressive goat, assess the herd as a whole. Can you tell how this goat's behavior is impacting the other goats? Do you see a shift in the social hierarchy of your herd? Sometimes, the solution to the problem of an overly aggressive goat is to remove the animal from your herd and sell the goat to another farm. You don't want one goat creating stress for the entire herd.

The Bully Goat

As social animals, goats naturally establish a social hierarchy within their herds. This social structure—a pecking order, to borrow a term from the henhouse—helps maintain order, reduce conflict, and set clear ranks within the herd. Most of the time, good-natured goats accept their role within the group without question, but from time to time, goats will vie for higher positions within the herd by bullying the others.

Photo Courtesy of Gayle Ewen

A bully goat will push another goat away from the food or water. You may observe the bully blocking access to these areas by simply positioning its body in the way, or the bully goat may physically shove or headbutt its target. When you visit the goat barn, the bully will keep other goats away so it can have all your attention for itself. The bully's target, or the low-ranking goat in the hierarchy, may exhibit submissive behavior. You might see it moving out of the other goat's way, stepping back from feed or water dishes, and avoiding direct eye contact with other goats.

As a hobby goat farmer, you do not have the power to eliminate the social hierarchy of your goat herd. You can, however, do a few things that might shake up the pecking order, which might put an end to the bully goat's reign. Removing a goat from the herd or introducing a new one into the group can change the dynamics. You can also make sure the housing and pasture space is roomy enough for all the goats to live harmoniously together. Close quarters can lead to conflicts within the herd.

Take the time to learn the hierarchy of your goat herd and identify the alpha goat, as well as the low-ranking members of the herd. Be on the lookout for bullying behavior by monitoring your herd and observing their interactions with each other. It is possible to put an end to bullying, but hobby farmers need to first have a good understanding of the instinctive behaviors of goats.

Dealing with Difficult Births

Another common challenge for small-scale hobby goat farmers involves the kidding process. The majority of the time, does will deliver their kids without help, just as nature intended. But sometimes, things don't go as planned. If a doe is having difficulty delivering her kid, prompt intervention is essential to ensure the safety and well-being of both the goat and her kids.

Prior to the goat's labor and delivery, you can take steps to increase the likelihood of a smooth delivery, including feeding the doe the correct amount of food so she gets the nutrients she needs, offering her access to plenty of fresh, clean water, monitoring her overall health, and practicing good hygiene in the goat barn. When the doe goes into labor, watch her closely for signs of trouble.

If the goat labors for more than an hour and has not made any progress toward delivering her kid, it might mean the baby is in an abnormal position, such as breech, which is causing the delivery to stall. It is possible for you to reposition the kid by inserting a clean, lubricated, gloved hand into the birth canal and gently moving the kid; however, if you are inexperienced at doing this, it is best to seek medical help. You could inadvertently cause injuries to the doe and the kid if you are not sure what you are doing.

Call your veterinarian if the goat appears to be in severe distress and is having difficulty delivering her kids. Veterinarians have the training, expertise, and equipment needed to address complex birthing issues. In the meantime, you can provide comfort and support to the doe. Keep her calm and comfortable, offer fresh water, and keep her birthing stall quiet and free of distractions.

Occasionally, the doe's labor and delivery will be uneventful, yet the kid experiences some struggles as soon as it is born. Be prepared for such emergencies by having a birthing kit ready. Your birthing kit should include clean towels, iodine solution for cleaning the kid's umbilical area, scissors, and other supplies. Once the kid is born, clean its nose and mouth to clear away any mucus. Contact your veterinarian right away if the kid does not respond, is lethargic, or appears to have breathing difficulties.

Effective Strategies for Goat Training and Handling

Goats and humans have shared a special bond for thousands of years, and although goats can be stubborn and mischievous, they are also intelligent, trainable, and genuinely want to please their humans. The earlier you can start handling and socializing your goats, the easier they will be to manage and train as they get older. Training and handling goats require diligence, patience, and an understanding of their natural behavior. It may be challenging at times, but there are plenty of strategies that can help.

Young goats are more eager to learn and receptive to training. Start by playing with the kids when they are very young so they get used to you, begin to trust you, and learn to accept you handling them. Use positive reinforcement to train your goats. A special treat, like apple slices, attention and praise, and extra time being brushed and stroked, will encourage goats to repeat the behavior you want them to learn. Never train your goats using negative reinforcement, like punishments or hitting them. Study after study has proven that negative reinforcement is counterintuitive to learning.

Be consistent in your training, and make sure all members of your family are using the same cues and commands as you. You will confuse and frustrate the goats if you give them mixed signals. You should also avoid becoming frustrated. Training animals takes time, and if you become frustrated, the goats can sense it. The underlying anxiety will be detrimental to their learning. Being consistent also means that you should handle your goats on a regular basis so they become accustomed to human interaction, and you learn to recognize when the goats have had enough of their training for the day.

Photo Courtesy of
Katelynn Sharp

If you plan to show your goats or you want to welcome visitors to your hobby farm, your goats will need to learn how to walk on a lead and stop on command. These are also helpful commands to use when you are leading your goat to the milking stand or when the veterinarian wants to examine it. You can have lesson time in your pasture to get them used to the rope lead and commands. Keep the lessons short, but repeat them often.

Summary

Not every moment with your goats will be sunshine and rainbows. Every so often, you will encounter a problem or challenge that you will need to work through. As a new hobby goat farmer, it can be easy to blame yourself when a problem arises but take comfort in knowing that you are not alone. Other small-scale goat farmers have experienced the same issues as you. Use them as a resource for troubleshooting, and learn from their experiences. This will make your problem-solving more effective and help you connect with other hobby goat farmers in your area.

CHAPTER 9

Milk and Dairy Goat Management

Many small-scale goat farmers are most attracted to hobby farming for the milk the goats produce. To be honest, that was high on our "Pros" list when we were considering starting our own goat herd. Producing our own fresh milk, cheese, yogurt, and other dairy products was important to us. The more I learned about goat milk, the more I wanted to have access to fresh goat milk for my family. I have never been a fan of cow's milk. Drinking cow's milk upsets my digestive system, and I don't like the fact that commercial cow's milk contains hormones. Goat milk, on the other hand, offers some potential benefits.

I was fascinated to learn that more people across the globe drink goat's milk on a regular basis than cow's milk. Humans began drinking goat's milk long before cow's milk, according to anthropologists who have determined that goats were first domesticated in the Fertile Crescent as far back as 10,000 BC.

Goat's milk has a different protein structure from goat's milk, and some people find it easier to digest. Individuals who are allergic to casein, the protein in cow's milk, may find that goat's milk does not cause the same allergic reactions. Additionally, goat's milk contains less lactose than cow's milk, making it potentially more tolerable for people with lactose sensitivities.

You may have heard that goat's milk is naturally homogenized, and that's why it is easier to digest. That's a bit misleading. You know how raw cow's milk separates and the cream rises to the surface? Goat's milk has smaller fat globules that remain suspended in the milk for a longer period, so this happens more slowly, and the layer of cream is thinner. The milk, however, is not naturally homogenized. Like cow's milk,

Photo Courtesy of
Eva Jordan

commercially sold goat's milk undergoes a mechanical homogenization process to break up the fat globules and make the milk more stable. The only difference is that goat's milk does not require as much homogenization as cow's milk because the fat globules are already smaller.

Goat's milk is a good source of nutrients. It contains calcium, phosphorus, potassium, vitamin A, and vitamin D. When compared to cow's milk, goat's milk is higher in vitamin B6 and vitamin A. Goat's milk also contains bioactive compounds that support good gut health.

Many people contend that goat's milk is beneficial for skin health because it contains anti-inflammatory and moisturizing properties. Plenty of hobby goat farmers have created lucrative side businesses by using goat's milk to make soaps, lotions, and other skin-care products.

If you are like me, all these potential benefits of goat's milk will have you eager to start milking your goats. However, before you do, you need to have a thorough understanding of the milking process, the equipment needed, the safe handling and storage of milk, and how you can best utilize the milk. All of these topics, and more, will be covered in this chapter.

Understanding the Milking Process

Goats are mammals, and if you recall your elementary school science lessons, you know that the ability to produce milk to feed their young is one of the main characteristics of all mammals, along with the presence of fur or hair, sweat glands, and a few other important things. Milk production only starts when the female's pregnancy hormones trigger the mammary glands to lactate. In simpler terms, the goat has to give birth before she will produce milk.

Once a doe has kidded, she can be milked. Her body will continue to produce milk as long as it thinks it is needed. In nature, the nursing kid stimulates the doe's body to keep up the milk production, and when the kid is weaned, the milk production will slow down and stop. The act of milking a goat also encourages the doe to continue producing milk. Since there is no weaning kid to trigger her body to reduce milk output, does that mean the doe will produce milk forever? No, that's not how it works. Lactation is hard on does. Most goat farmers only milk them for a recommended time, after which they will allow them to go dry and have a period of rest; then, they will breed them again. That standard time frame is 305 days. Studies have shown that 305 days is the optimal time for the health of the doe.

Hand Milking Versus Milking Equipment for Small-Scale Herds

Hobby farmers who are interested in starting a small goat herd will have many decisions to make, including how they will milk their goats—by hand or with a milking machine. Let's look at the pros and cons of each method as it pertains to a small-scale goat farm.

One advantage of hand milking is that it is practically free. You don't need to invest in costly equipment. All you need are your hands, a bucket, and some soap. Hand milking your goats gives you an opportunity to bond with them and build trust. It is also quieter and less stressful than machine milking. Another advantage is that hand milking is not tied to electricity or to the location where the milking machine is housed. You can milk your goats anywhere you want.

Hand milking goats can be both time-consuming and labor-intensive, which are big drawbacks for some hobby farmers. Depending on your experience and the individual goat, it may take you between seven and fifteen minutes to milk each goat. If you have six goats, for example, count on spending an hour and a half each morning and another hour and a half each evening doing the milking. It is not an easy task either. I found hand milking to be tiring. At the risk of sounding like a princess, milking made my hands hurt and cramp. It takes some physical effort to milk goats. Lastly, when you hand milk your goats, you have to be extra careful to prevent contamination.

A milking machine, on the other hand, can significantly speed up the milking process. If you have several goats and a day job, investing in a milking machine can free up a lot of your time and effort. It is less physically demanding and labor-intensive to use a milking machine. Another advantage of milking machines is that they provide a more consistent milking process, reducing the risk of human error and ensuring all the milk has been extracted. Today's milking machines have been designed with an emphasis on good hygiene and sanitation. Milking this way can reduce contamination and increase the quality of the milk.

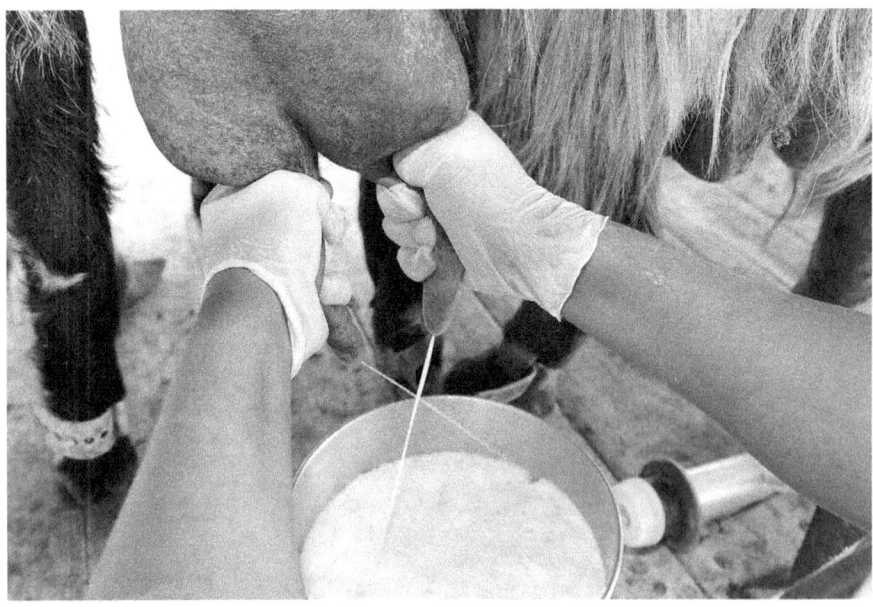

Perhaps the biggest drawback to using milking machines is the cost. Not only are the machines themselves expensive to purchase, but they need routine maintenance and occasional repairs. You will also need a dedicated, sanitary place to house the milking machine, like a milking parlor, and you need electricity to run it. This brings up another downside—the reliance on electricity and technology. During power outages or technical failures, the milking process grinds to a halt, requiring goat farmers to hand milk their herd as a plan B. Lastly, using a machine to milk your goats is impersonal. Your opportunity to interact and bond with your goats will be gone.

Ultimately, the decision to hand milk your herd or use a milking machine depends on your personal situation, time, budget, the size of your herd, and your own preferences.

Maintaining a Milking Schedule

To maximize their milk output, your goats should be milked on a twice-a-day schedule. Since goats can't tell time, you can set the milking schedule to fit into your schedule, but there are a few caveats to that. First, once you set a milking schedule, you have to stick to it every day, including on your days off work. What that means is that if you do the first milking of the day at 6:00 a.m., you will need to do the second one at 6:00 p.m. If you like to sleep in every morning and you don't do the first milking of the day until noon, you will find yourself in the barn at midnight for the second milking.

Typically, goat farmers will plan their milking schedule to include a morning milking and an evening milking. The 6:00 a.m. and 6:00 p.m. plan is a good one. It takes advantage of the goat's morning milk—milk that accumulated overnight—which is often higher in fat content and more suitable for cheesemaking. Morning milkings also relieve the doe's udder that has filled up overnight. The second milking of the day empties the udder once again, promotes continued milk production, and relieves the discomfort of a full udder.

Maintaining a consistent milking schedule is the key to a consistent milk supply. Goats are creatures of habit. They thrive on routine and

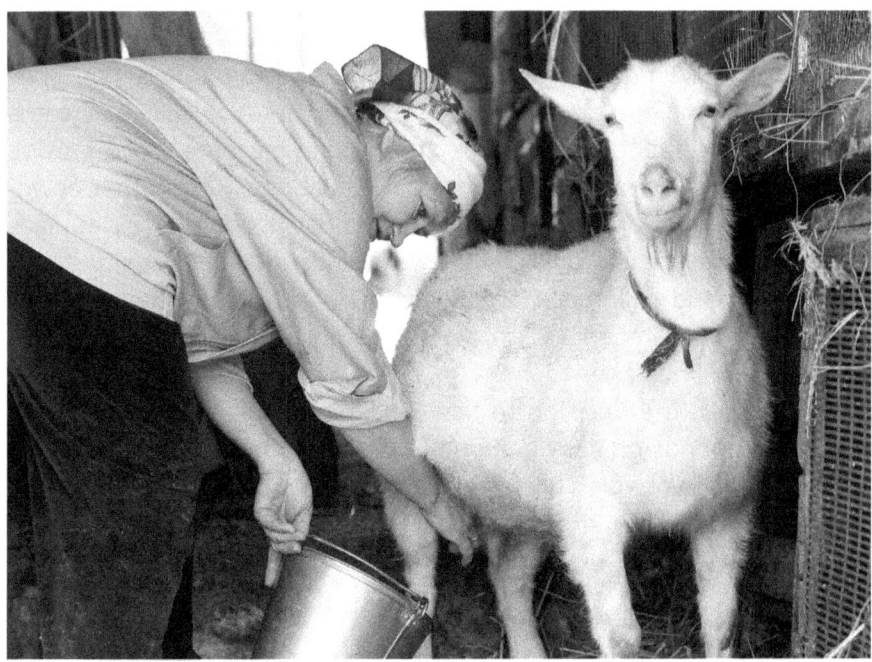

predictability. A consistent milking schedule gives them a sense of security and reduces their stress level. The result is happier, healthier goats. Scheduled milkings help prevent udder engorgement, which can be painful for the does. It can also cause inflammation and mastitis, an infection of the udder. When goats are milked on a regular basis, there is a reduced risk of bacterial growth and contamination in the udder, giving you higher milk quality. An irregular milking schedule can confuse the goat's natural lactation cycle and make its body think that it is weaning time.

Goats might love a consistent schedule, but for us humans, that can pose some challenges. There might be demands on your time that prevent you from sticking to this schedule. You may have to work late or go on a business trip. You may have dinner plans one night, a wedding to attend, or a family vacation. You could get sick and need to stay in bed, or your car could break down on the way home from work. All of these things have happened to me, and they prevented me from milking my goats on time. My advice would be to have a backup. Don't be the only one in your family who knows how to milk the goats. Teach a trusted friend, neighbor, coworker, or extended family member how to do it. Have your goat sitter on call. In fact, have a few people you can call in an emergency to help get the milking done on time.

Milking Equipment

Before you dive into milking your small-scale goat herd, you need to have some equipment on hand. The list of items you need will vary, depending on your milking method. You may not need everything on this list, but this can serve as a guide when purchasing milking equipment for your hobby farm.

Milk Pail – You need a pail or bucket to collect the milk, but not just any bucket will do. You need a new, clean, food-grade pail, preferably one with no seams. Bits of milk can collect in the seams, spoil, and taint the flavor of the milk. Your milk pail should be made of metal as a pail made of plastic, a porous material, can absorb milk particles and become funky smelling.

Milk Stand or Stanchion – This is a device that holds the goat in place so you can milk her. Most of them are elevated, making it easier for you to access the udder. They typically also have a feed container or hay pouch to keep the goat occupied and distracted while she is being milked.

Milk Stool – Save your back. Get a stool to sit on while you are hand-milking your goats so you aren't bending over for a long time.

Cleaning Supplies – The goat's udder should be thoroughly washed before you milk her, so you will need udder wash detergent or a mild sanitizing solution, as well as some clean, dry cloths or paper towels. If you are using cloth towels, you will need to wash them in hot water after each use.

Strainer or Filter – When milking goats, place a stainless-steel strainer or filter over the opening of the pail to collect hair, dirt, or debris that might fall as you milk. You can purchase disposable, one-use filters or use a metal strainer. The stainless-steel strainer will need to be washed and sanitized after each use.

Airtight Storage Containers – After the milk has been collected, you will need to store it in the refrigerator. You can reuse plastic milk jugs from the grocery store if you want a low-cost, recyclable option. Personally, I do not like to use plastic containers to store milk. I much prefer glass containers. I have a collection of glass jugs with narrow mouths and those rubber-sealed, clamp-on stoppers.

Labels – Depending on the number of goats you have, you could end up with a fridge full of milk bottles from different dates and goats. It can be easy to mix them all up. You will need labels, markers, or some other system on which to record the date and goat's name on each bottle. Those glass jugs that I use—I can write right on them with a marker, then wash it off when I need to.

Milking Machine – If you opt to use a milking machine, you will need to have the various items that are required for that, such as teat cups and a vacuum pump. The machine will likely have a receiving system that includes a milk line that carries the milk from the teat cups to a collection container. Have extra parts and supplies for the milking machine on hand at all times in case they are needed.

Milk Machine Sanitation Supplies – To properly clean and disinfect the milking machine after each use, you will need to have cleaning brushes, sanitizing solution, and other disinfecting equipment.

FUN FACT
Rain, Rain, Go Away

Goats are relatively easygoing in most situations, but there's one thing they strongly dislike: rain. If your goats suddenly seek shelter, you might be in for a rainstorm. Most goats will find the nearest shelter whenever it rains, sometimes even seeking refuge before rain starts to fall. In addition to their dislike of rain, many goats abhor mud and puddles. For this reason, farmers should provide their goats with plenty of shelter from the rain and dry areas where their hooves won't sink into the mud.

Always follow the directions and recommendations that come with the milking machine.

Udder Balm – Udder balm and teat dip are antiseptic, moisturizing ointments that are used to protect goats' udders. These products are intended to keep the skin moisturized, healthy, and soft. They relieve skin irritation, cracking, and chapping. Milking, either by hand or by machine, is tough on goats. Udder balm soothes soreness and dry skin.

Recordkeeping System – It is important to keep track of information related to milk output, so you need to have a recordkeeping system that works for you. Some hobby farmers keep paper records, like a journal or notebook. This information can also be put onto a computer spreadsheet or a cellphone app. In addition to milk yield, you can include your observations about your goats' health, diet, and other data.

Safe Milk Handling and Storage

How you handle, store, and process goat milk requires you to pay careful attention to hygiene. Milk is susceptible to contamination, spoilage, and bacterial growth that can render it inedible or, worse, can sicken people who drink it. To ensure the cleanliness, safety, and quality of the milk, you need to take a number of steps to manage the milk you collect. They include these four activities:

Washing Your Hands – The first line of defense against germs and bacteria should always be handwashing. Before milking your goats and handling the milk, thoroughly wash your hands with antibacterial soap and warm water. Wash your hands often throughout the process.

Use Food-Grade Collection Devices – You can take another step to protect the safety of your goat milk by using only food-grade milk pails, buckets, collection containers, and storage containers. Stainless steel milk pails and collection containers and glass storage containers are preferred, as these materials are not porous, and milk particles cannot penetrate the surface.

Cool the Milk Immediately – As soon as you finish milking your goats, cool the milk to preserve its freshness and quality. Ideally, you should cool the milk to below 40 degrees Fahrenheit within one hour of milking. One way to do this is to place the sealed container in a vat of ice water to quickly lower the milk's temperature before you store it in the fridge with the temperature set between 34 and 38 degrees Fahrenheit.

Clean Everything Often – After every use, thoroughly wash and sanitize all items that come in contact with milk. That includes the buckets, stanchion, collection containers, the milking machine, strainer, and more. Use the appropriate cleaning agents and sanitizers, and follow the recommended cleaning procedures for the milking machine.

The Pros and Cons of Pasteurizing Goat Milk

Raw goat's milk does not have a very long shelf life, even if you immediately store it in the refrigerator. If you are not planning to consume it right away, you can consider pasteurizing it to remove harmful pathogens and increase its longevity. However, some people prefer to keep their goat milk unpasteurized because the pasteurization process alters the flavor of the milk and decreases its nutrient levels. Let's take a moment to discuss the pros and cons of pasteurizing your goat milk.

Raw milk can harbor harmful bacteria, including E. coli, salmonella, and listeria. Pasteurization effectively destroys harmful pathogens, such as bacteria, parasites, and viruses, thus reducing the risk of foodborne illnesses. For people with weakened immune systems, consuming unpasteurized milk can be especially dangerous. The pasteurization process extends the shelf life of goat's milk by killing the bacteria that cause the milk to spoil. It stays fresher longer than raw milk. As we will see momentarily, many places require goat's milk to be pasteurized before it is sold.

Many individuals can taste the difference between pasteurized and unpasteurized goat's milk and argue that the pasteurization process takes away from the taste of the milk. Because the process involves heating the

*Photo Courtesy of
Eva Jordan*

milk, some of the beneficial nutrients and enzymes are lost along with the harmful pathogens. The nutritional value of pasteurized goat's milk also declines as the process destroys the natural probiotics—beneficial microorganisms that offer health benefits and support good gut health. In fact, the unique microbial properties in raw goat milk are important for some artisanal dairy products, like goat cheese and yogurt. Pasteurizing goat's milk takes time and energy, which may be in short supply on a hobby farm. Lastly, there is the cost factor. For small-scale goat farmers, the investment in pasteurization equipment may not be practical.

The Pasteurization Process

If you decide to pasteurize your goat's milk, either for your own family's consumption or because you wish to sell it, you will need to familiarize yourself with the process, be committed to cleanliness and the proper milk-handling protocol, and have all the necessary equipment.

To begin the pasteurization process, thoroughly wash and sanitize all the equipment you will be using, including the pot, thermometer, and stirring utensils. Then, take the fresh, chilled goat's milk that you have collected and stored in your refrigerator and pour it into a large

stainless-steel pot or double boiler. Place the pot on the stove and set the heat to medium.

Using a food thermometer, closely monitor the temperature of the milk throughout the heating process. The milk should be gradually heated to 145 degrees Fahrenheit. The milk needs to remain at this temperature for 30 minutes. By gently stirring the milk often during this time, you will ensure that the temperature is evenly distributed and prevent the milk from scorching.

After 30 minutes, remove the milk from the heat. The next step requires the milk to be quickly brought back down to 40 degrees Fahrenheit within 20 to 30 minutes, and this cannot be accomplished by simply putting the pot of milk in the refrigerator. You will need to speed up the process by placing the pot of milk into an ice-water bath. Stir the milk constantly to cool all the hot spots and ensure a consistent temperature. Use your food thermometer to check the temperature.

Only after the milk has been sufficiently cooled can you pour it into freshly washed, sanitized bottles. The bottles should be tightly sealed to keep air out and stored in a refrigerator that has been set to between 34 and 38 degrees Fahrenheit. Pasteurized goat's milk should remain fresh in the refrigerator for seven to ten days, but always check the milk for signs of spoilage before you drink it. Spoiled milk smells off and may have an unusual appearance.

The Legalities of Selling Raw Goat's Milk

As a food item, goat's milk is subject to laws and regulations regarding the sale of this product. The U.S. Food and Drug Administration (USFDA) has banned the sale of raw goat's milk across state lines but leaves in-state sale requirements and regulations up to each state. The USFDA has also set recommended guidelines for ensuring the safety of raw milk sales. These regulations only apply to the sale of goat milk. The milk you keep for you and your family's consumption does not fall under these rules.

At the state level, the laws vary regarding the sale of raw goat's milk. A few states allow goat farmers to sell unpasteurized milk for human consumption, while others only permit milk sold for pet consumption or to make cosmetic products. In many states, the sale of raw goat's milk is prohibited. In addition, some states require goat farmers, even small-scale operations, to obtain permits or licenses to sell raw milk. As part of the permit approval, the farm is subject to health inspections and regular product testing. The regulations might dictate what information must be on the product's label and what, if any, warnings should be included.

It is important to do your own research to obtain the current rules and to ask questions so that you understand the specific regulations and guidelines in your state before you start selling raw goat's milk from your hobby farm. You don't want to risk getting into legal trouble, being hit with hefty fines, or being liable if the unpasteurized milk from your farm sickens one of your customers. Engage with your state's health department, agriculture department, and any other relevant regulatory agency to help you navigate the legal requirements and make sure you are in compliance. Consult with legal experts as well to confirm that you are operating within the law.

Exploring Dairy Product Options and Value-Added Opportunities

There's money to be made from goat's milk—if you can find the right niche for your business. This is common among hobby farmers. They use the products produced on their hobby farm as a revenue source to help offset farm expenses. The milk that your goats produce can be used for a number of things, not just for a beverage to accompany your chocolate chip cookies or something to pour over your cereal. Goat's milk is versatile and can be used in a number of ways.

Goat Milk for Soap Making

Goat milk is an ideal ingredient for soap-making due to its potential benefits for the skin. In fact, goat milk soap is often considered a luxury product. Goat milk is rich in natural fats and proteins that help nourish the skin as it moisturizes. Those fats also contribute to the creamy, thick lather of the soap. The milk contains lactic acid, a natural alpha hydroxy acid known for its exfoliating properties. It can be beneficial for removing dead skin cells, smoothing skin, and giving a fresher appearance as it cleanses.

Goat milk also contains vitamins A, D, and B6, as well as other vital minerals like calcium and selenium. These and other components in the milk have been reported to have anti-inflammatory properties and are better for people with sensitive skin.

When making soap, some people will use goat milk as the liquid base and add it to the lye solution. Others prefer to use frozen or powdered goat milk as it reduces the chance of scorching or overheating. Soap makers often enhance goat milk soap with additional natural ingredients, such as honey, oatmeal, botanicals, clays, essential oils, or herbs, to further boost the soap's nourishing properties. Like other handmade soaps, goat milk soap needs to cure for a certain period before it is ready to use.

Goat Milk for Cheese Making

Goat milk has a unique composition and flavor that makes it ideally suited for cheese making. Goat cheese is widely considered to be an artisan or upscale product that is favored by chefs and foodies. Cheese made from goat milk can range in taste from mild and creamy to bold and tangy, depending on the cheese-making recipe, the breed of goat, and even the goat's diet.

When compared to cow's milk, goat milk has a higher proportion of medium-chain fatty acids. This translates into a creamier texture. Additionally, the protein structure of the milk can contribute to the formation of a firmer curd during the cheese-making process. Goat milk has a higher natural acidity compared to cow's milk. This can be a plus in cheese making as it facilitates the coagulation of proteins and the full development of the flavors.

The smaller-sized fat globules in goat's milk can make goat cheese more digestible for people who have a sensitivity to cow's milk.

A wide variety of cheeses can be made using goat milk, including soft cheeses like feta, cream cheese, and chevre and aged cheeses like cheddar, Gouda, and blue cheese. Some cheesemakers choose to use pasteurized goat milk in their cheese-making, while others prefer to work with raw milk to take advantage of the natural enzymes and flavors. During the cheese-making process, the goat milk is treated with rennet and starter cultures to promote coagulation in the same way that cow's milk is treated.

The milk forms curds, which are then cut, heated, and drained to separate the curds from the whey. The curds are processed even further, according to the recipe. Soft, spreadable cheeses, for example, are drained and lightly pressed. Hard cheeses are pressed for a longer time to expel more whey and create a compact texture. Aging the cheese imparts unique flavors, textures, and characteristics to it. Some goat cheeses are aged for only a short time, while others are aged for months or even years.

During the aging process, the cheese's surface may develop natural molds and form a rind that contributes to both its flavor and appearance. Proper care and storage of the cheese during the aging process are essential to give the cheese the desired flavor and texture, as are temperature control, humidity maintenance, and regular turning.

Goat Milk for Cosmetics

When using goat milk in cosmetic products, it's vital to source high-quality, natural goat milk. Goat milk skincare and cosmetic products pair well with other items produced on hobby farms, such as lavender, honey, and mint. Many hobby farmers use goat milk to make lotions and moisturizers, bath bombs, bath salts, lip balm, and more, which they sell from their farm stand, at farmers' markets, and online.

Goat Milk for the Pharmaceutical Industry

The pharmaceutical industry has taken notice of the unique properties of goat milk. Its bioactive compounds and ease of digestion make it a viable alternative to cow's milk as an ingredient in medicines. There has been some promising research regarding the potential pharmaceutical applications of goat milk in studies and clinical trials. In the near future, this could become a lucrative avenue for small-scale goat farmers looking for customers to buy their milk.

Summary

The desire for goat milk is the reason why many hobby farms get goats in the first place. Milking your goats, either by hand or with a milking machine, can give you and your family fresh milk for drinking or to be made into various products. Cleanliness and attention to hygiene are vitally important before, during, and after the milking process to ensure the safety of the milk. Hobby goat farmers have the option to pasteurize their milk to kill potentially harmful pathogens or leave the milk in the raw state. If they plan to sell their goat milk, however, they need to know the regulations in their state regarding the sale of milk. Many states have banned or greatly restricted the sale of raw, unpasteurized goat milk.

Goat milk can be used to make cheese or other dairy products, like butter, yogurt, and ice cream. It can also be made into soap, skincare and cosmetic products, and more.

CHAPTER 10

Meat Goat Production and Marketing

The goats I've raised on my hobby farm have been mostly used for milk production and for 4-H projects for my kids. Many of the 4-H goats on our farm have been meat goats that my daughters raised, exhibited at our local fair, and sold at the livestock auction. Admittedly, I have not participated in the butchering process, nor do I eat goat meat (vegetarian here!), but I do have some experience raising meat goats.

Vegetarianism aside, I am not alone when I say I have never eaten goat meat. As a whole, consumers in the United States are lagging behind in this area. Goat meat is widely consumed by people around the world, especially in regions of Africa, Asia, the Middle East, and the Caribbean. In fact, it is a staple in the traditional dishes and cultural cuisine of many places, yet goat meat, called chevron, has historically been less popular than other meats like beef, pork, and chicken among Americans. Several factors contribute to this, but the tide seems to be turning. American attitudes about goat meat are gradually changing, which is good news for small-scale goat farmers.

In general, North America was settled by mostly European immigrants rather than immigrants from regions where goat

DID YOU KNOW

Most Popular Protein

Goat is the most popular meat in the world, consumed by 60 to 75 percent of the global population. Egypt has the highest production of goat meat, followed by Morocco, Jordan, Tunisia, and Lebanon. Despite a lack of widespread popularity in America, goat meat is a high-protein, vitamin-packed option with a lower barrier to entry than cattle ranching. In fact, 65 percent of all meat consumed worldwide is goat meat.

meat was ingrained in the culture. During pioneer days, the majority of meat consumption revolved around beef, poultry, and pork. Goats were not commercially raised like cattle were, making it difficult for consumers to find goat meat in mainstream grocery stores. The lack of availability of chevron made it difficult for people to make goat meat part of their everyday diets.

Another factor could be that goat meat production requires specialized farming techniques. As browsers, goats prefer different types of vegetation than cattle, posing challenges for large-scale goat farmers. Lastly, many Americans have historically not been aware of the nutritional benefits and potential of goat meat.

In recent years, there have been noticeable shifts in attitudes about chevron, indicating that interest in goat meat has increased in the U.S. The American food scene is evolving as more people are seeking new cuisines and ingredients. Goat meat's unique flavor and versatility make it appealing to foodies interested in exploring diverse flavors and traditional cuisine from other cultures. The rise of ethnic cuisines—spurred on by well-known celebrity chefs—has moved chevron dishes into the spotlight, introduced them to a wider audience, and banished old misinformation about goat meat.

Consumers are more conscious of their food choices these days and are seeking sustainable and ethical meat options. Goat meat's lower environmental impact is attractive to people looking for more earth-friendly food choices. Goat meat is part of the farm-to-fork move-ment that encourages people to look for locally-grown food options. As a result, small-scale goat farms that specialize in meat production are gaining traction.

Lastly, goat meat is better for you than beef. Chevron is lower in fat, making it a good lean protein source for health-conscious consumers.

Goat Meat Considerations

For small-scale goat farmers, the opportunity to develop a niche mar-ket offering antibiotic-free or organic goat meat to specialty shops and local farm-to-fork restaurants is greater than ever before. But before you jump into meat goat farming, there are some things to consider.

Select the Right Breed – While it is possible to eat all goat breeds, some breeds are better suited for meat production. These breeds have the right combination of physical characteristics, genetic factors, and growth potential to make them more efficient at producing quality meat. Breeds like Boer, Kiko, and Spanish goats have more muscular builds.

Pasture and Housing Space – Meat goats need to have adequate shelter and sizable pasture space. A well-built goat barn with good venti-lation and floor drainage will protect the animals from the elements and predators and allow the goats to live in conditions that resist the spread of disease. A spacious pasture allows the goats to get plenty of exercise, which helps build and tone their muscles, burn excess fat, and produce meat that is lean and tasty.

Nutrition – The old adage, "You are what you eat," applies to meat goats, too. You simply cannot expect to produce high-quality meat if you are feeding your goats a subpar diet. Feed your herd of meat goats a balanced diet of top-quality food that meets their unique nutritional requirements. Work with your veterinarian or a livestock nutritionist to develop a feeding program that includes high-quality grains, forages, and, if necessary, supplements. You will also need to be committed to good pasture management and practice pasture rotation to prevent overgrazing and promote healthy pasture growth. If this is a struggle on your hobby farm, seek expert help to ensure that your pasture provides a variety of forage types for balanced nutrition.

Budget Considerations – To be successful as a small-scale meat goat farmer, you have to have the financial resources available to invest in your herd before you can hope to see a profit from the sale of chevron. In addition to housing and pasture space, you will need to purchase the best food available and arrange for routine veterinary care for your goats. Establish a health-care plan with your veterinarian that includes check-ups, disease prevention, parasite control, and routine vaccinations. Any goat that becomes sick or injured will need immediate veterinary care. All of this adds up. Expect a lot of expenses before you see any income.

Managing Meat Goat Growth and Weight Gain

We use the expression "fattening a goat for market," but this is really an unfortunate and misleading phrase. Although you want your meat goats to bulk up before slaughter, you don't want them to be fat. Fat goats produce fatty meat, which is less flavorful and of a poorer quality. By practicing meat goat growth and weight gain management, you can control the rate of gain of your meat goats and, quite literally, trim the fat. Here are a few ways to accomplish this.

Feed for Protein and Energy – Humans wanting to lose weight and build muscle often try to drastically reduce their carb intake. Meat goats, however, need an adequate amount of carbs in their diet. In fact, it is the energy from carbohydrates that fuels the animal's growth rate and weight performance. Meat goats need the right balance of both carbo-hydrates and protein for muscle growth and overall health. Remember that the digestive system of goats is quite different from that of humans. Don't fall into the trap of applying methods that work for human body-builders—less carbs and more protein—and expect the same results with your goats.

Keep Thorough Records – Monitor the diets, weight, and rate of growth of the meat goats in your herd and keep thorough records. As a newcomer to hobby farm meat goat raising, you can't expect to have all the answers. The more data you can collect, however, the better equipped you will be to make informed decisions to improve the growth and quality of your animals. In your records, make note of the breeding stock of your goats, weight at birth, health check-ups, and what food they eat. After a few years, you'll be able to spot trends or patterns that allow you to tweak some of the variables to improve the meat quality.

Regular Exercise – Exercise is key for building muscle mass. When goats work their muscles, they burn off excess fat, leaving them with strong, toned muscles. Goats will get some regular exercise when they browse around the pasture, but there are a few things you can do to encourage your meat goats to work out more. You can, for example, walk them every day. My kids did this with their show goats in the months leading up to the county fair. It not only forces the goats to move more, but it trains them to walk on a lead and obey commands. You can also fill your pasture with toys to entice your goats to work out. A big wooden

spool, a few old tires, a teeter-totter, and—if you are handy with a saw and some nails—a wooden ramp or elevated walkway will keep your meat goats' minds and bodies active.

Set Achievable Weight-Gain Goals – Instead of, as the saying implies, trying to rapidly fatten your meat goats right before they go to slaughter, it is best to set goals for consistent weight gain over a long period of time. Each individual goat varies, of course, depending on its breed and genetic makeup, but experienced meat goat farmers typically aim for a rate of gain of between one-half and one-third pounds per day. To achieve this, the goat will need to consume between three and a half and four pounds of additional feed per day. Once the goat has stopped growing and has peaked at its maximum growth, back off the additional feed. It no longer needs the extra boost to spur muscle growth, and, at this point, it will only cause the animal to gain fat.

Slim Down Overweight Goats – If you practice good weight gain management from an early age, your meat goats shouldn't become over-weight. However, it does happen from time to time. Your goats could, for example, be eating too much of a good thing growing in your pasture, or the neighborhood children could be sneaking unhealthy treats to your goats on a regular basis. Perhaps one dominant goat is eating more than its fair share of the grain. It happens.

Fortunately, there are ways to help an overweight goat slim down and improve its muscle tone ahead of slaughtering time. First, decrease the amount of feed the goat has on a daily basis. Goats need to eat between 1.5 and 2% of their body weight to maintain good health; however, most hobby farmers feed their goats much more than that— as much as 3 to 5%. You can safely decrease the goat's food intake without compromising its health. If you suspect the goat is overeating in the pasture, you can muzzle it to curb excessive eating. Just be sure the animal can still drink water with the muzzle on. Get the goat exercising more so the fat can be converted to muscle. Two or three times a day, walk the goat around the pasture, or better yet, get it to run. Weight loss in goats doesn't happen overnight, but if you are diligent, the animal will be able to shed the extra pounds and present at market with an improved body mass.

Slaughtering and Meat Processing Options

When meat goats are at their ideal size and body mass, it is butchering time. For many small-scale hobby farmers, this is an emotional time. After all, you have spent time caring for and working with your goats, learning their personalities, and bonding with them. It is natural to experience a range of feelings when butchering time rolls around. But keep in mind that your purpose in raising these goats is for their meat. It is a necessary part of farming. That said, hobby farmers have a few decisions to make when it comes to slaughtering their goats and processing the meat. The first decision is, do you want to do the job yourself or send your goats to a local slaughterhouse? There are pros and cons to each.

On-Farm Slaughtering

When you opt to slaughter your own meat goats on your hobby farm, you have complete control over the entire process. You have raised your goats in clean, well-kept conditions and ensured that they were treated ethically. When you keep the butchering process on your farm and under your control, you can be sure that the animals are slaughtered under humane conditions to minimize stress and suffering. Not only will this result in better-tasting, higher-quality meat, but you will feel better knowing that your animals were slaughtered quickly and efficiently in a clean, hygienic environment.

Butchering your own animals will save you money as you will not need to pay for the services of a local slaughterhouse. On-site butchering is also appealing to hobby farmers who want to be as self-reliant as possible. It is part of the sustainable lifestyle that many hobby farmers seek.

On the other hand, slaughtering animals on your farm can be emotionally challenging. Many people find it distressing to take a life, even if it is for food production. You will need to be prepared for the realities of the slaughtering process, which can be brutal, messy, smelly, and heart-wrenching.

Butchering also requires a certain level of knowledge and skill. You want to slaughter the animals in the most humane and ethical way possible, yet this takes some practice to perfect. An inexperienced person could inadvertently cause the goat to suffer, contaminate the meat, or create unnecessary waste. It is also time-consuming and physically demanding. You need to be sure you are up for the task.

You will also need to have all the necessary equipment on hand, have a sterile place to do the butchering, and pay careful attention to food safety protocol. You will need an on-site facility in which to do the butchering and meat processing. The room should have running water, a floor drain to remove blood, and be constructed using easy-to-clean materials, like ceramic tiles. It can be quite costly to have such a facility constructed and to have plumbing added. You will also need sharp knives, bone saws, bowls and dishes, plastic wrap, and butcher paper. Cleanliness and hygiene should be a top concern to reduce the risk of contamination and prevent the spread of diseases.

There may be legal restrictions to consider as well. In some communities, livestock slaughtering on residential property is subject to various local, state, and even federal regulations. If a hobby farmer fails to comply with these regulations, he or she can face legal consequences. Some of these legal regulations involve the disposal of waste materials, like bones, entrails, hide, and other byproducts of the slaughtering process. Hobby farmers need to work within the legal parameters to ensure the proper disposal of byproducts to prevent environmental issues.

Off-Site Slaughtering

When you use a local or regional slaughterhouse to butcher and process your meat goats, you can be assured that the facility is regularly inspected by health department officials whose job it is to make sure the slaughterhouse is operating within the ethical and hygiene laws set forth by the federal government. The employees at the slaughterhouse are experienced and know how to butcher goats quickly and efficiently, with little waste and suffering.

Using a slaughterhouse allows hobby farmers to distance themselves from the brutal and emotional process. They can say goodbye to their goats in their own driveway when they are loaded into a trailer and not

participate in the gruesome butchering itself. You can save yourself some heartache—as well as time and energy.

Slaughterhouses have guidelines in place to ensure proper cuts and meat quality. The end result is meat that is more uniform and consistent, which is what consumers demand in the marketplace.

Because slaughterhouses are in the business of butchering animals, they have all the necessary licenses in place to comply with local, state, and federal regulations regarding meat processing. Your local ordinances may prohibit you from butchering animals on your farm, but that is not a worry at the slaughterhouse. They have a system in place to dispose of animal byproducts.

There are, however, a few drawbacks to using a local slaughterhouse to butcher your meat goats. The first is cost. Using a professional slaughterhouse involves paying a fee for the slaughter, processing, and butchering services. There will be a transportation fee if the slaughterhouse picks up your goats and delivers them to the slaughterhouse.

When you send your goats to a slaughterhouse, you are giving up control over the meat production process. For hobby farmers who prioritize self-reliance and transparency, this might be concerning. If one of the goals of your hobby farm is to be sustainable and independent, this is a step in the wrong direction. You will no longer have a say in how your goats are treated and how the meat is processed.

Scheduling time at a busy slaughterhouse can sometimes be a logistical nightmare. Depending on the time of year and other demands of the slaughterhouse, you may need to book your time well in advance and hope that your goats will be ready when the time comes. You may need to coordinate transportation and arrange to pick up the packaged meat afterward. None of this is guaranteed to fit neatly into your schedule. You will, instead, be at the mercy of the slaughterhouse's time frame.

Packaging Goat Meat

As a meat goat farmer, even on a small scale, you should be familiar with the different cuts of chevron and how they are typically used. If you are packaging the goat meat yourself, this is particularly important, but having a good working knowledge of goat meat cuts will help you when you market your meat to your customers. Here are some of the common cuts of chevron:

Shoulder – The shoulder can be cut into shoulder roasts or shoulder chops. The roasts can be slow-cooked, braised, or slow-roasted. The chops, cross-section slices from the shoulder, are usually prepared by grilling, broiling, or stewing.

Ribs – The ribs, or rack of goat, can be roasted whole or divided into individual ribs for roasting, barbecuing, or grilling. The rib chops, individual rib portions, are ideal for pan frying, oven baking, or grilling.

Hindquarters – The goat's hindquarters can be cut into leg roasts, leg steaks, or leg chops. Leg roast, considered one of the more flavorful chevron cuts, can either be cooked whole or divided into smaller roasts. Leg steaks are thick slices from the leg, while leg chops more closely resemble lamb chops. Both of these cuts can be broiled, pan-fried, or grilled.

Cuts Of Goat

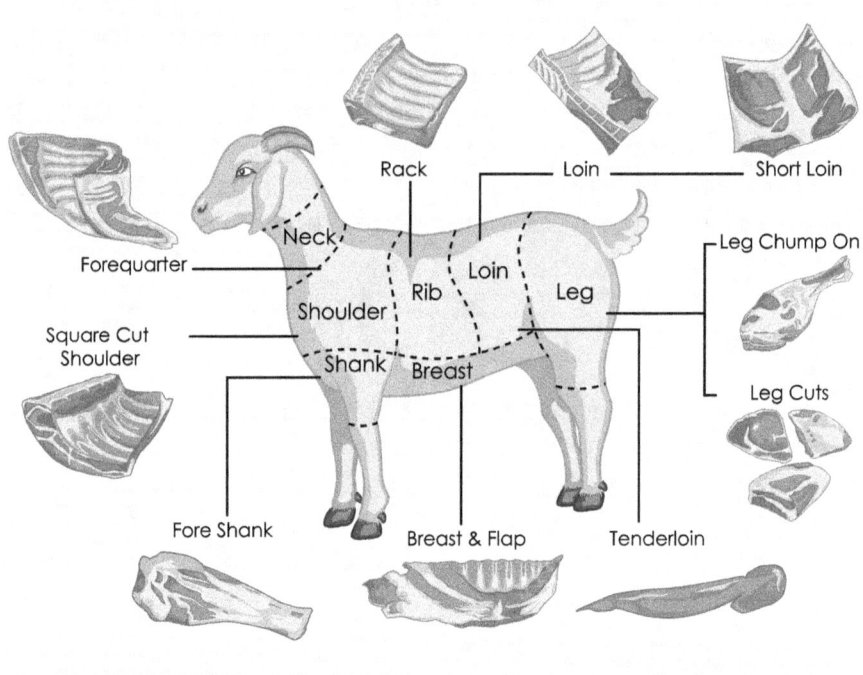

Shank – The shank cut is often a bit tougher cut of chevron; however, it becomes tender and tastier when it is cooked over low heat for a long time.

Loin – The loin area can be processed into loin chops or loin roasts. The tender cut of the loin area, loin chops are suitable for broiling, grilling, and pan searing. Loin roasts are often roasted whole to bring out their flavor and increase their tenderness.

Flank – Flank steaks are typically thin cuts that can be sliced for stir-fries, sandwiches, or fajitas. They are great marinated, too.

Organ Meat – The goat's liver is packed with nutrients. People enjoy it fried, grilled, or made into pate. The heart and stomach, or tripe, are traditionally used in stews. In the traditional cuisine of some cultures, kidney meat is a favorite.

Storing Goat Meat

Once the chevron is processed, you will need to safely store the meat, either for your own family's consumption, to sell directly from your farm, or to take it to a local market. You will need a large, dedicated freezer that is located in a convenient place. You will need to regularly clean and sanitize the interior of the freezer to ensure the integrity of the meat. It is also not a bad idea to have a generator hooked up to the freezer in case of power outages. How devastating it would be to have all your time, money, and hard work ruined because of a power outage.

Selling Goat Meat

There is an increasing demand for goat meat, especially in culturally diverse communities where chevron is a staple in traditional dishes. It is also gaining in popularity in areas where consumers value locally sourced, organic, sustainable food options. This all means there are more business opportunities for small-scale hobby farm meat goat producers than ever before. The key to a successful side business selling chevron is to find the right customer base and develop marketing strategies to increase awareness about your products.

Legalities of Selling Goat Meat

The legalities of selling chevron vary from location to location, so it is important to do your homework and understand the regulations in your area before you get involved in meat goat farming. The regulations may require your meat and your farm to undergo inspections by governmental agencies to ensure that health and safety standards are being met. This will likely involve inspections of your livestock and their living conditions, inspections of your slaughtering facilities, and testing of the final products. You may be required to apply for a license or permit before you can sell your goat meat. It may include a food handling permit, a meat processing permit, and a business license.

All meat that is sold is required to have proper labeling to inform customers about the product's origin, processing, and potential allergens. As part of the health department's disease control and prevention protocol, you will also need to have a system in place to trace the sources of your meat. There may be forms and documents that you need to maintain.

You may discover that there are regulations governing how the chevron is stored, as well as how it is transported. This includes temperature-controlled or refrigerated delivery vehicles.

The places where you sell the chevron may also be the target of regulations. If you are selling the meat from your hobby farm, you may have a different set of regulations and requirements than if you sold your products at a farmers' market, in a retail store or specialty shop, or online. Lastly, depending on your location, there will probably be tax implications for selling chevron.

Who Is Your Ideal Customer?

Before you set a marketing plan and work to attract customers who are interested in purchasing your chevron, you first need to pin down your ideal customer. This will vary depending on where you live. In general, however, goat meat customers fall into one of three categories. Let's take a moment to explore these.

Cultural and Ethnic Markets

Goat meat is an important staple in many cultures. Not only is it an ingredient in favorite meals, including Indian curry dishes and Mexican *birria*, but goat meat plays a role in cultural holidays and celebrations. Chevron is also consumed during religious celebrations for people of the

Muslim and Jewish faiths. If you live in or near an area with a diverse ethnic makeup, you can promote your product to groups who historically seek out goat meat, especially ahead of cultural celebrations and holidays.

Sustainable Living Markets

As people become more aware of how foods are produced on a large-scale, commercial basis, many of them have made the conscious decision to focus on organic, locally grown foods that are produced at small-scale or hobby farms. For these people, access to fresh, local, ethically raised meat is important because it aligns with their dual goals of eating healthier, chemical-free foods and not contributing to the environmental impact of traditional, large-scale agricultural practices. Reach out to these people at farmers' markets and local food stands, as they are potential customers for your chevron.

Foodies

In the last several years, chefs and foodies have discovered the rich flavor and versatility of chevron. A lean protein with a distinctive taste, the meat is at the forefront of the foodie movement as chefs and food aficionados alike look for recipes and ingredients that are new to them, packed with flavor, and give their dishes an unexpected, exotic flair. To market your chevron to foodies, visit restaurants, specialty markets, and cooking classes that are known for experimenting with flavors from around the world.

Where to Sell Goat Meat

Another consideration to discuss is where you plan to sell your goat meat. There are more outlets through which you can sell your chevron than you may think, but deciding on which ones to use will depend on your location, the local demand for chevron, and your personal goals for your hobby farm. Here are a few options.

On-Farm Sales

Perhaps you envision selling your goat meat directly to local customers from your hobby farm. That is a viable option, as many small-scale goat farmers use this method. You can keep things small and informal—a customer knocks on your door and buys a few packages of meat from your freezer—or more businesslike, with a dedicated farm store, set hours, and a salesperson.

Farmers' Markets

One of the benefits of setting up a booth at a local farmers' market is that you will be joining other sellers who are also offering locally produced agricultural products. The larger variety of vendors means more people who regularly visit the venue. The type of people who frequent farmers' markets fit your ideal customer profile.

Ethnic Markets and Specialty Shops

If you live in or near a community with a diverse population, there may be established ethnic markets or specialty shops that cater to the same clientele that you are seeking. It may be possible for you to either sell your meat to the store or place your products in the store on a commission basis.

Local Restaurants and Chefs

Are there restaurants or caterers in your area that specialize in international cuisines? Reach out to the restaurant owner, chef, or culinary professional to discuss the possibility of sourcing your locally raised goat meat for their menu options. With so many farm-to-fork-themed restaurants, managers and chefs are looking to build relationships with local farmers who can provide them with a consistently high-quality product.

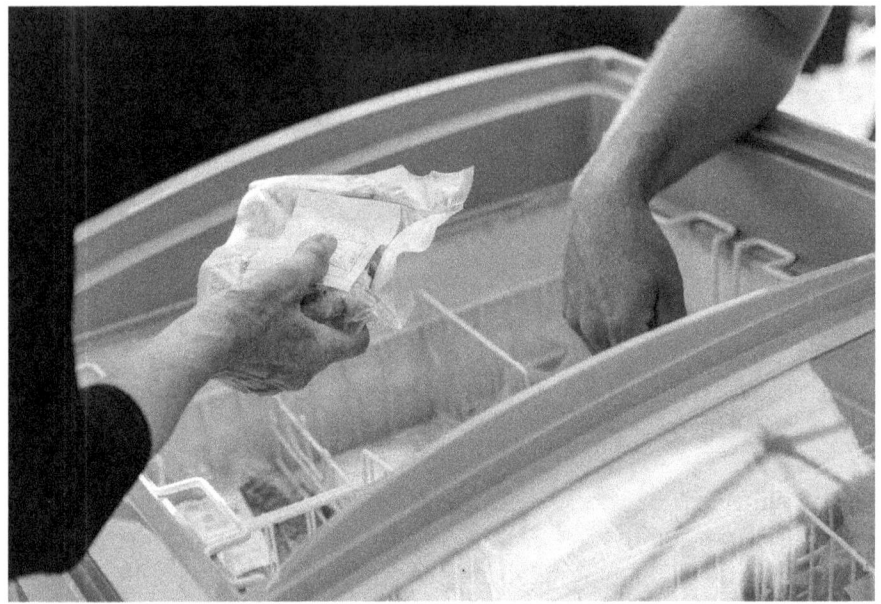

Food Co-ops

Food co-ops are food distribution methods that use the collective power of members to buy food in bulk and divide it among the members, who all benefit from the cost savings. Oftentimes, local farmers partic- ipate in food co-ops. It is a way for you to get your product to a wider customer base by selling to one entity.

Online

Yes, it is possible to sell goat meat online. You can either set up your own website to market and sell your product or join an e-commerce plat- form that specializes in local, organic, or ethnic food. When selling your product online, however, you will need to work through the logistics of shipping the meat and complying with food safety regulations.

Marketing Your Goat Meat

Thanks to the internet and social media, it is now possible to set up a low-cost marketing plan that reaches your target customers. You can, for example, promote your product on Facebook, Facebook Marketplace, or Instagram, all for free. You can use your social media accounts to not only sell your goat meat but also educate your followers about your hobby farm, your commitment to sustainability and ethical agriculture practices, and the nutritional benefits of goat meat. It is possible to host your own website without a big cost commitment and without a background in web design. There are several web hosting companies that have low rates and user-friendly templates to help individuals and small businesses create their own websites. Lastly, you can design and print an informative flyer that you can give out at farmers' markets and other venues. The flyer should point out the benefits of goat meat and include information about your hobby farm, as well as your contact information. You can have these professionally printed or simply print them off from your home computer.

Summary

If you plan to raise meat goats on your hobby farm, there are a number of things to consider. First, you need to make sure that you have the space, financial resources, and time to raise your goats to produce high-quality meat. You need to carefully manage the food intake of your goats to make sure they are getting the nutrients they need to build muscle but aren't overeating and growing fat. You will need to decide if you want to butcher your goats on your farm or send them to a slaughterhouse. Lastly, you will need to determine who your ideal customer is, where to find them, and how to market your chevron to them. Raising meat goats can be an enjoyable experience and a money-making side business as long as you are committed to the process.

CHAPTER 11

Long-Term Goat Farm Planning and Sustainability

A few years after buying our small hobby farm, we dipped our toes in the water of goat farming by welcoming a pair of goats to our farm. We had no plan in place for raising goats and just sort of learned as we went—making plenty of mistakes along the way. It wasn't until our goat barn was destroyed and we went back to square one that we actually made a long-term plan for our hobby farm. I would advise you to be more thoughtful and proactive than we were. We could have saved ourselves some time, money, energy, and stress if we had a long-term plan and a sustainability guideline in place from the start.

What does it mean to set a long-term plan? How do you establish a guideline for sustainability on a small-scale goat farm? These are both excellent questions we will tackle throughout this chapter.

What Does Long-Term Planning Involve?

When we were forced to rebuild our goat barn, we finally did what we should have done much earlier—think about what we wanted to achieve with our goat farm. We had to ask ourselves the same questions from Chapter 1. Did we want milk goats or meat goats? Did we want to turn a profit from our goat farm? And so on. The answers to these questions led to other questions and decisions that pointed us in the direction we wanted to go. Here are some things to consider:

Financial Investment

Raising goats, even on a small scale, takes money. It can be an expensive hobby, which is why many hobby farmers look for ways to offset the cost by producing sellable goods on their farms. As part of your long-term planning, create a detailed budget of goat-related expenses, then divide it down by the number of goats you currently have to give you a per-goat estimate of expenses. This is a helpful number to know when you are thinking about expanding your goat herd.

Time Investment

Raising goats is a big-time commitment. There are chores that need to be done multiple times a day, every single day. When setting your long-term plans, you need to be realistic about your available time, both now and in the future. There was a family in our community who raised goats. Their daughters showed against my daughters in the county fair—and they always beat my daughters! But after a few years, they sold all their goats and shut down their hobby farm. What happened? Well, it had to do with time. For several years, the mom, who handled the bulk of the farm work, was a stay-at-home parent while taking a few college classes per semester. Their daughters were young. But then, the mother graduated with her nursing degree and took a demanding third-shift job at a nearby hospital, and the daughters got involved in middle school and high school sports. The demands on their time had changed.

Growth Opportunities

If your goals for your hobby farm involve growth and expansion, you need to work that into your long-term plan. Knowing that you plan to expand your herd and having a timeline for growth will guide you when making decisions about pasture space, goat barn size, and other infrastructure.

The Big Picture

Think about how your goat herd fits into the big picture of your hobby farm. If your hobby farm is diverse, goats may be just a small portion of what you do. Your farm might also include chickens, rabbits, ducks, or other livestock animals, along with vegetables and fruits. Such diversity is beneficial for families trying to grow their own food. It also offers another revenue stream if your long-term goal is to develop your hobby farm into a profitable business.

If you keep the overall goal in mind, you can plan the best way to integrate goats into your plan. For example, goats and goat products can complement other marketable goods produced on your hobby farm. Lavender, herbs, and honey can be used to make goat cheese and goat milk skincare items. A small-scale market selling goat meat would attract more customers if it also sold poultry meat, rabbit meat, turkey meat, and so on.

What Does Sustainable Farming Mean?

For small-scale goat farmers on a hobby farm, sustainability refers to the practice of managing the farm in a manner that balances financial, environmental, and social concerns. The purpose or goal of sustainable farming is to ensure the well-being of the farm, your family, and the surrounding ecosystem. Since hobby farms operate at such a small level compared to commercial farms, achieving sustainability is oftentimes easier, although the basic principles remain the same.

Sustainability means being good stewards of the environment. It is about implementing strategies to minimize your farm's impact on the land, such as preventing soil erosion, water pollution, and habitat loss. Sustainable practices involve adding more native plants to replace invasive species, taking steps to improve the biodiversity of wildlife, and using an integrated and targeted pest management system. When you are mindful of waste and conserve resources, you are working within the framework of sustainability.

For you and your family, sustainability means producing a portion of the food you consume to reduce your reliance on outside sources,

like grocery stores, but it also includes enjoying the physical and mental benefits of working on the farm, raising your own food, and connecting with nature. The concept of living a sustainable lifestyle is also linked to preserving and passing down traditional farming skills and practices to the next generation.

A sustainable hobby goat farm is also committed to ensuring the well-being of the goat herd. It is more than just providing proper housing, nutritious food, and access to clean water—although these are also important. It means providing opportunities for the goats to engage in their natural, instinctual behaviors, such as browsing and climbing.

Sustainability, Responsible Breeding, and Culling

Responsible breeding practices are essential for sustainability on a hobby farm. It will ensure the health of the animals, maintain the integrity of the breed, and improve the overall population of farm goats. When you

practice responsible breeding on your small-scale farm, you are making the commitment to produce healthy, genetically sound, and genetically diverse animals that are free of hereditary diseases and disorders.

Unfortunately, many small hobby farmers do not follow careful protocols when breeding their goats. Yes, the end result is more goats on their farm, but the quality suffers. This could impact the reputation of your hobby farm. If you are selling inferior goats, subpar goat meat, and poor-quality goat products, it will reflect poorly on your farm as a whole. Word will get around, and your customer base will decrease.

Sadly, hobby goat farmers should also have a culling plan in place. Not every goat born on your farm will be a top-notch animal that is the picture of health. Some will be born prematurely or have defects. Some will be sickly or will fall short of hitting the breed standard. A goat that has a chronic health problem may be cute, and you may have a soft spot in your heart for underdogs—or undergoats—however, keeping a goat like this is a drain on your resources and could jeopardize the health of your whole herd.

Culling sickly or poor-quality goats allows you to keep the quality and productivity of your herd at a high standard. As part of your sustainable breeding program, you will need to come up with a plan for culling goats as needed. Humanely euthanizing an unsatisfactory goat means you can allocate your resources toward healthy, productive animals.

Your culling plan should include guidelines or parameters for determining which goats need to be culled and why. It must also establish your method of culling and how you plan to dispose of the carcass. You should also regularly review your culling plan and make adjustments as necessary to make sure your plan falls in line with your long-term goals for your hobby goat farm.

Recordkeeping and Data Management

The sustainability goals and long-term plans for your small-scale hobby goat farm should also include a system for recordkeeping. Even small goat herds can benefit from accurate and organized records. Don't dismiss this task as unnecessary, time-consuming, or tedious. The information you record will help you analyze trends so you can make informed decisions regarding your goat farm.

Your farm records should include health information, such as illnesses, vaccination and deworming schedules, and routine medical care. They should also record breeding information, such as breeding dates, parentage, and breeding outcomes. This data will help you maintain genetic diversity in your herd and ensure effective animal husbandry.

Depending on the purpose of your goat farm—milk, meat, or fiber—your recordkeeping system should track productivity. Milk yields, fiber output, or meat weights for each goat should be recorded so you can identify the most productive animals in your herd. More importantly, it can help you to determine why some goats are more productive than others. You may be able to tweak a few things to increase the productivity of the rest of your goats based on the information you glean from your records.

Include costs and expenses in your records as well. If you are profiting from your hobby farm as a business enterprise, you will have a business ledger for tracking financial information, but it is also a good idea to keep notes of costs and expenses in your records. You'll be able to easily see if your feed costs are increasing or if you are spending so much on fencing repairs that it makes more sense to replace it. This portion of

your records can also include an inventory of supplies, from milk buckets to bales of hay and straw.

Recordkeeping might look like old-school notebooks or journals or like technology-based spreadsheets and computer software programs. Just choose the method that is easiest and least stressful for you. Having complete and accurate records gives you a way to discover patterns, trends, and performances over time. Armed with this information, you will be better equipped to make informed decisions, set goals, and make long-term plans for your hobby farm.

Expanding or Diversifying Your Small-Scale Goat Farming Enterprise

Throughout this book, we have focused mostly on goat milk and goat meat as money-making ventures for small-scale goat farmers. There are, of course, other avenues to explore if you are looking to expand or diversify your hobby farm enterprise. Some of these can even dovetail into your existing goat endeavors.

Textile

We briefly touched on raising goats for the textile industry. Depending on the breed of goat you get, it is possible to get fiber and meat from the same animals. Valais goats, for example, can provide you with fiber in addition to quality meat. These goats do not produce as much wool as other goat breeds, but their dual role makes them attractive to hobby farmers looking to branch out into other areas.

Agro-Tourism

Agro-tourism is a growing trend in local tourism. Small farms add seasonal or year-round attractions to go along with the agricultural practices they currently engage in. These local attractions offer a way for people

to learn about farming practices, see farm animals, and experience farm life, if only for an afternoon.

Some hobby farms open their gates year-round and give their guests an authentic farm experience. Visitors can watch farm activities and even participate in some farm chores as a way to learn about rural life and food production. Others are only open seasonally, like pumpkin patches, you-pick apple orchards, lavender farms, or sunflower fields. The more attractions at these types of agro-tourism attractions, the more visitors you will attract and the more repeat customers you will gain. Goats are a big draw at places like this. Young and old alike enjoy petting the goats, feeding them some grain, and watching their funny antics.

Rent-A-Goat, Part One

Even if you do not have your own apple orchard or pumpkin patch, it is possible to partner with a farmer who does and arrange for them to "rent" a few of your goats for the duration of their agro-tourism season. A pumpkin farmer, for example, may want to attract more visitors by adding a petting zoo portion to his farm just for the six weeks he is open to visitors every year. It is easier and less costly for him to rent a few goats for a petting zoo than it would be for him to raise goats year-round.

Rent-A-Goat, Part Two

Small-scale goat farmers can also rent out their goats to clear over-grown areas or to keep weeds and brush from getting out of control. Brush-control goats are becoming increasingly popular for several rea-sons. First, goats are an environmentally friendly alternative to gas or electric-powered mowing equipment. Goats are able to reach areas that may be too difficult to access with large equipment, like a steep hillside or rock-strewn area. Goat herds are hired to clear underbrush

Photo Courtesy of
Gayle Ewen

in wildfire-prone places, for example. In my county, the power company routinely hires goats to tackle the plant growth at their solar farm. State and national parks often utilize goats to help them with their plant removal needs.

Goat Yoga

A unique and increasingly popular form of yoga, goat yoga combines traditional yoga practice with the presence of playful, friendly goats. The participants engage in an outdoor yoga class while goats wander around them. It gets fun when the goats join in. They might climb on people, frolic around them, or nuzzle them. Yoga is intended to relieve stress and increase mindful relaxation, and when you combine it with goats, the mood is light, entertaining, and fun. Of course, the goats are always good for Instagram-worthy photos that are comical and sharable. You could partner with a yoga instructor to host regular or occasional goat yoga sessions at your hobby farm.

Summary

Practicing sustainability on your hobby goat farm involves being a good steward of the land, avoiding wasteful activities, and conserving resources. It also includes engaging in responsible breeding protocols, culling unhealthy or defective animals from your goat herd, and setting long-term goals for your hobby farm to help keep you on track. Keeping good, thorough records will help you gather the data you need over time, which will help guide your decision-making. A long-term plan may include diversifying your hobby farm to add other potentially profitable endeavors, such as agro-tourism attractions, which will help your farm grow in different directions.

FUN FACT
Mitochondrial DNA

The first breed of domesticated goat most likely descended from the wild bezoar ibex goat. However, scientists studying the origins of domestic goats have recently discovered six distinct strains of mitochondrial DNA. This data helps scientists understand where early goat domestication occurred and how goats have spread worldwide. These separate strains of mitochondrial DNA suggest that goats may have been domesticated at different locations around the same time. Surviving archaeological evidence of early goat domestication is being investigated at sites including Turkey, Syria, Israel, and Jordan. Today there are over 300 distinct breeds of domestic goats worldwide.

CHAPTER 12

Goat Health Emergency Preparedness

Back in Chapter 4, I promised to share one of our more bizarre and, for my husband at least, embarrassing goat emergencies. I will preface the story by saying that Juju the goat came out of the experience virtually unscathed, except for a few scratches and bruises. It started one weekend afternoon when my husband—the assistant fire chief for our small community's volunteer fire department—went out to do some yard work. I thought it was odd that I didn't hear the lawn mower, but I went about my day. About a half hour later, he came inside and said he needed my help.

I live on an old farm that was established in the 1880s. Out back, there is an old windmill that once pumped water for the livestock animals. The windmill is still standing, but the pump workings are no longer there. There is a deep well beneath the windmill that was covered with some plywood years ago. It was hidden amid weeds and vines growing up the sides of the windmill.

Two of our goats, Juju and Dina, escaped their pasture

DID YOU KNOW
World's Oldest Goat

The world's oldest goat was named McGinty and lived an impressive 22 years and five months. She resided in Haling Island, Hampshire, United Kingdom, and was owned by Doris C. Long. McGinty died in November 2003, but her exceptional genes were passed on because her granddaughter, a goat named Daisy, lived for 19 years. The average life span of a domestic goat is between 8 and 12 years, depending on breed, genetics, and living conditions.

and wandered over to the windmill to nibble on the vines. Their combined weight was too much for the old, rotting plywood. It gave way, plunging first Juju, then Dina into the well. We had no idea when this happened, but my husband heard them crying for help when he went outside. He finally found them in the narrow well, one on top of the other! He was able to get Dina out without assistance, as she was on top of Juju. But poor Juju was fur-ther down and, by this time, exhausted.

Together, my husband and I tried everything we could to get Juju out of the well. We put a ladder down the shaft in hopes that Juju would climb out herself. My husband tried to go down into the hole to coax her up the ladder. Of course, Juju is not skilled at ladder climbing, and the well shaft was too narrow for my husband to get in and push her up. In fact, when he tried to go down close to her, Juju panicked. I think she was feeling claustrophobic. And I know she was exhausted, stressed, and thirsty. We were able to lower a bucket of water down, and she eagerly drank it.

An hour or more passed, and it became clear that we would not be able to get Juju out of the well without assistance. My husband—remember, he is the assistant fire chief—got his fire radio and prepared to call dispatch to summon help to our farm. As I made "Lassie, get help! Timmy is stuck in the well!" jokes, my husband desperately tried to think of another way—any other way—so he wouldn't have to call the fire department for help.

Within a minute, the tones of his pager dropped, and the dispatcher read off our address. "Animal stuck in a well. Repeat, animal stuck in a well." Several fire trucks loaded with my husband's firefighting bud-dies soon pulled into our driveway. Firefighters, at least the ones in our department, are dedicated, hardworking, and brave. They are also a bunch of jokesters who never miss an opportunity to razz each other. The good-natured bantering started immediately, to my husband's deep embarrassment. The firefighters set up a hoist over the well opening, and all nominated the newest member of the crew, a young, skinny 18-year-old named Brandon, to descend the ladder and harness Juju. "Probie initiation," they called it.

Poor Juju had no strength left and was stressed about the whole situation. She allowed Brandon to squeeze in next to her and tighten the straps, but she seemed to know she was the butt of all the jokes and ribbing. When the fire chief told Brandon if he got any more intimate with the goat, he would have to marry her, Juju let out a surprised bleat.

Juju was unceremoniously hoisted out of the well to the cheers of the firefighters. We were pleased to see that she did not appear to have any severe injuries from her ordeal. She had a slight limp as we walked her back to the barn, but she bounced back within a few hours. The vet came to check her out the next day and found no broken bones.

That year, at the annual firefighters' award dinner, both my husband and Brandon were awarded the "Golden Goat" Award. And the firefighters still make goat noises at my husband, especially when they want to break the tension and get a guaranteed laugh.

Emergencies

A hobby farm is not immune to emergency situations. In addition to the Golden Goat incident, we had, as I have mentioned, a tree come down on the goat barn during a bad storm. You may experience similar unexpected events on your farm, either weather-related, medical-related, or some other incident. Putting an emergency preparedness plan together will allow you to handle any curveball thrown your way. Let's look at some of the possible emergency scenarios that could arise and discuss what your emergency plan should look like.

Weather Event

Weather disasters can happen at any time. A tornado, flooding, wild-fire, lightning, or blizzard could totally disrupt operations at your hobby farm. Do you have a place to house your goat herd if their goat barn is damaged due to a storm? We didn't. Our goats spent a few hours in the garage while we frantically made some calls. Connect with other hobby goat farmers in your area and develop a backup plan in case someone needs emergency temporary housing. You should also have a plan and the necessary equipment to deal with extreme cold or extremely hot temperatures in a safe and effective way.

Evacuation

Something might happen that forces you, your family, and the other living creatures on your hobby farm to evacuate. It could be a weather event, like a hurricane, flood, or wildfire. Or it could be something else, like a man-made disaster. There was a train derailment in a neighboring community a few years ago, and one of the rail cars leaked a hazardous

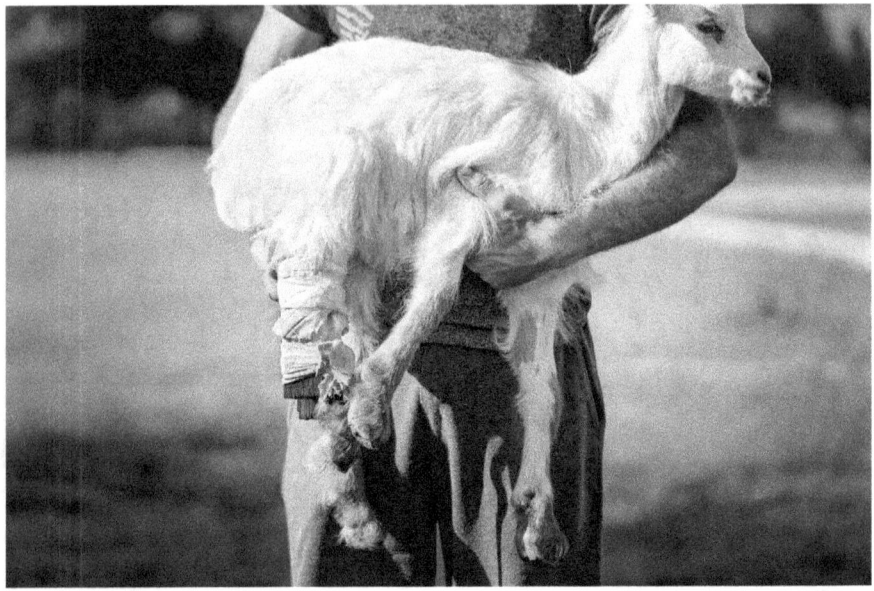

liquid. The authorities evacuated all the houses within a mile radius of the spill. A goat-farming family we know was one of the homes that was evacuated. They quickly loaded up their small goat herd into their horse trailer and moved them out of the area until they were cleared to go back.

Injuries

As curious and mischievous as goats are, it is not out of the question for injuries to occur. A goat could cut itself on fencing, hurt a hoof climbing on a rock, or scratch an eye while browsing for leafy forage. As a goat farmer, your task is to identify potential risks to mitigate possible injuries. Fix fence wires that are sticking out, make sure climbing materials are stable and secure, and so on.

Disease Outbreak

A contagious disease outbreak is an emergency situation because it can quickly spread through your whole herd and sicken all your goats. The best way to prepare for this type of emergency is to take steps to prevent it from happening. It starts with establishing biosecurity protocols. New goats coming onto your farm should be isolated for a few weeks to make sure they are not carrying a disease. Each hobby farm should have an isolation area well away from the goat housing. There, new animals can be temporarily kept, and sick goats can be separated from the rest of the herd. Biosecurity measures go beyond quarantining. They include limiting visitors to your farm, properly sanitizing equipment, and washing your hands before and after handling your goats. They also mean culling goats that are chronically ill or contaminated with a deadly disease.

Power Outage

Your goats won't care if the power goes out—they don't need electric lights. You do, however, especially if you rely on a freezer to store your goat meat, a milking machine to milk your dairy goats, and a refrigerator to keep the goat milk from spoiling. As part of your emergency preparedness plan, you should invest in a generator to power the equipment you need for your goat farm operation.

Your Absence

Teach everyone in your family how to take care of your goats. Even train a few friends or neighbors so there are several people who are knowledgeable about how to do your goat chores. If you are suddenly called away from your goat farm, there will be people who can step in and tend to your animals. You may have a family emergency that takes you away from the farm, or you may have a business trip to take. You may experience your own medical emergency that sends you to the hospital for a few days. Rather than having the added stress of finding a temporary goat keeper during this time, you will already have trained people at the ready.

An Emergency Kit

Create an emergency kit with all the supplies you need to handle a problem should it arise. We have our emergency kit in a large plastic tote with a tightly fitting lid. In it, we have a first aid kit, medications, wound-care products like bandages and antiseptic, electrolyte formulas, a few gallons of water, a bag of hay, and a container of feed. We also have some collars and leads to help us handle the goats if needed. A pair of scissors, some wire cutters, a goat brush, some hand sanitizer, antibacterial wipes, and some old towels are also in our emergency kit—just in case.

Taped to the inside of the lid of our emergency kit is a list of phone numbers and contact information. The same list is on our fridge in the house, too. Our veterinarian's name and phone number are at the top of the list, followed by the emergency after-hours phone number. The list also includes the local animal control department and the county extension office. A few of our goat farming friends who are knowledgeable and experienced in goat first aid and emergency care are also on the lists, along with their contact numbers. Lastly, the list has the names and numbers of a few friends and neighbors who have been trained to do our goat chores—people we can call at the last minute if we need help.

Identifying and Responding to Potential Outbreaks

One of the most devastating emergencies that could befall your hobby goat farm is a deadly disease outbreak. You may not be able to keep your goats in a protective bubble to completely isolate them from viruses, but you can take steps to identify and react quickly in the face of a potential outbreak. Preventing outbreaks, recognizing early signs of disease, and knowing how to respond should be part of your emergency preparedness plan.

Educate Yourself

Like human diseases, viruses and illnesses that infect goats change and evolve constantly. New strains develop, old strains resurface, and illnesses that were once confined to one geographic region can spread to others. Educate yourself about common diseases in your area. Talk to your veterinarian about this. Attend workshops on goat health management. Join an online forum or discussion board that shares reputable information about the future of goat diseases. The more you stay informed about current goat health concerns, the better prepared you will be to implement best-practice protocols, recognize a potential outbreak, and know how to respond.

Prevention

We discussed disease outbreaks in Chapter 5, but to recap, you can mitigate illnesses in your small-scale goat herd by quarantining new goats for two or three weeks before you introduce them into your herd to make sure they are free of disease. Practice good hygiene on your farm with an emphasis on maintaining a clean, well-sanitized barn, feeding area, and milking parlor. Washing your hands before and after handling your goats, wearing dedicated boots when doing chores, and restricting the number of visitors to your farm are additional biosecurity measures you should take.

Health Inspections

Working with your veterinarian, create a system for conducting regular health check-ups on your goats. If you see signs of illness, such as loss of appetite, nasal discharge, coughing, diarrhea, lethargy, or behavioral changes, contact the veterinarian immediately. Isolate all goats that are showing symptoms of illness as soon as possible. You may have caught the illness early enough that you can prevent it from spreading through your entire herd.

Culling

It is devastating and heartbreaking, but in some cases, all the animals in a herd must be culled to prevent the further spread of a virus and contain an outbreak. Your veterinarian will probably be the one to make this call. Like most hobby goat farmers, you have probably come to love your goats, and you want to be there for them in sickness and in health, yet it is more important to keep your focus on the big picture. If destroying one small herd of goats can help stop a disease from spreading and impacting a large geographic region, it is worth the sacrifice.

Report the Outbreak

If a deadly or potentially deadly disease infects your goat herd, you or your veterinarian must report it to the local and state health departments, your county extension office, and, depending on the disease, the

Centers for Disease Control. This will give the appropriate agencies the data they need to analyze outbreaks and take steps to contain them. As the 2020 COVID pandemic has shown us, some viruses have the ability to jump from animals to humans, with dangerous consequences.

Minimizing the Impact of Emergencies

Goats are creatures of habit. They like the status quo. When an emergency arises, your goats' routine will be disrupted, causing them stress and anxiety. When our goats had to temporarily stay at the neighbor's farm while we rebuilt the goat barn, the animals were well cared for, but they were out of their normal routine. Several of them had a noticeable drop in milk production during this time. A few of them refused to eat. Emergencies are stressful, even for animals. Do what you can to minimize the impact of emergencies on your hobby farm goats, and strive to get things back to normal as quickly as possible.

Summary

Emergency situations can happen anywhere, even on a scale-scale hobby goat farm. By taking the necessary steps to prepare for weather, health, or unexpected fluke incidents like a goat falling down a well, you will be better able to handle the curveballs that life sometimes throws at us. Having an emergency plan, backup housing, an emergency kit, and a list of contact numbers of people who can help will take much of the stress off you and reduce the anxiety your goats are experiencing. This holds true for disease outbreaks, too. Know what to look for and how to respond so you can quickly contain a potentially devastating outbreak.